HORSES

Horses

A 4,000-YEAR GENETIC JOURNEY
ACROSS THE WORLD

LUDOVIC ORLANDO

TRANSLATED BY
TERESA LAVENDER FAGAN

PRINCETON UNIVERSITY PRESS
PRINCETON & OXFORD

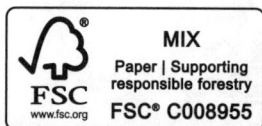

Original French edition: *La conquête du cheval: Une histoire génétique*.
Copyright © Odile Jacob, 2023.

Translation copyright © 2025 by Princeton University Press

Published by Princeton University Press
41 William Street, Princeton, New Jersey 08540
99 Banbury Road, Oxford OX2 6JX

press.princeton.edu

ISBN 9780691264127
ISBN (e-book) 9780691264134

British Library Cataloging-in-Publication Data is available

Editorial: Alison Kalett
Production Editorial: Natalie Baan
Jacket Design: Karl Spurzem
Production: Danielle Amatucci
Publicity: Kate Farquhar-Thomson and Matthew Taylor
Copyeditor: Anne Sanow

Jacket image: Seldaha Cosar / iStock

This book has been composed in Arno

Printed in the United States of America

10 9 8 7 6 5 4 3 2 1

To Andaine, for her patience in reading these pages in more or less completed versions, and her calm acceptance of the place horses have ultimately assumed in our lives.

To Pablo Librado, for being my traveling companion in the pursuit of the Pegasus Project dream.

CONTENTS

HORSES

1

Prologue

The Horse of the Past

Our species, *Homo sapiens*, spent most of its evolutionary history without the horse. But as soon as it appeared at our side, the entire history of the world was revolutionized. The world began to appear smaller than it ever had before—because what the horse offered us, above all, was a means to travel faster and farther than we ever could using just our legs. It provided an opportunity to explore the world, to meet with distant relations, to trade with them, to spread our languages, our cultures, and even our germs, at a speed that was previously unimaginable. The horse enabled us to connect with distant places and peoples in a relative blink of an eye, and to truly globalize ancient worlds for the first time. The horse also sometimes enabled us to interact and mix peacefully with others along the way, and other times—often, in fact— to wage wars of a different type, on horseback, or in chariots to which it was harnessed. It is indeed the horse that I wish to talk about here: the horse of kings and peasants, but also that of the nomads of the steppes, and polo players; the one that delights young and old alike, a delight that can at times cause people to lose their heads and complete fortunes by betting at horse races;

the one that carried the cavalry and its equipment, or the million and a half others that perished during the First World War; and the mustang, which became the uncontested emblem of the Great American West.

Through astonishing recent advances in DNA sequencing, we now know that the domestic horse appeared around 4,200 years ago in the steppes of the northern Caucasus region—in the lower valley of the Don and the Volga, to be more precise. That makes the horse one of the last large mammals to have been domesticated by humans, many thousands of years after the dog, the cow, the sheep, the goat, or the pig. It was "the most noble conquest" that humans have ever accomplished, the Count de Buffon wrote in his *Histoire naturelle,* a work of unparalleled popular success which in August 1753 would take him, the most illustrious naturalist of the Jardin Royal des Plantes, barely forty-five years old, from the doors of the French Academy of Sciences to those of the French Academy, the organization that defines the proper official use of the language in the country. He was the perfect incarnation of the Enlightenment.

The Horse in the History of Civilizations

The conquest of the horse may have been belated compared to other animals, but its success was immediate: in scarcely a few centuries, this new type of horse, henceforth domesticated, would be found quite far from its place of origin, from the coasts of the Atlantic to those of the Pacific; it was like an instrument of a new era that had just gotten underway. The German historian Reinhart Koselleck had the right idea when he proposed that the great periods in the history of our civilizations no longer be categorized by the metals that fashioned our weapons and tools—primarily the Copper Age, then the

Bronze Age, and finally the Iron Age—and suggested we espouse a dichotomy that is (in his opinion) not only simpler, but also more pertinent. One need only to determine whether humans already possessed the animal we are looking at here or not, to reveal a dynamic that had been little known but which was nonetheless fundamentally pivotal in human history. The proposal was to introduce an age before the horse, the Pre-Caballine Age (from the Latin name retained for the species, *Equus caballus*), and an age of the horse, the Caballine Age. This would emphasize just how profoundly the animal changed our societies and the great impact it truly had. Today we are living in a post-Caballine Age, the one after the horse, ever since combustion engines replaced the animal's motor force and caused the working horse itself to disappear from our cities and our countrysides.

But it was not so long ago that the horse was still an integral part of our daily universe, including in large cities such as Paris, London, or New York, where it pulled omnibuses that carried people from one point to another in those increasingly tentacular cities; but where it also brought supplies to businesses and stores, and also served in the maintenance of law and order, and even became a quintessential social marker in high society. At that time, everything—or almost—revolved around it. As proof of this: it took only a particularly devastating equine flu epidemic for the economy of one of the most powerful nations in the world, the United States, to grind to a halt. It was at the end of 1872 and the beginning of 1873, and according to various estimates, only between 2 percent and 18 percent of afflicted horses died. But close to three-quarters of them fell ill and developed a fever and wracking cough that made it impossible for them to move and thus to work for many long weeks. Without horsepower, that nation which earlier was as proud as it was thriving, and which nothing seemed to be able to daunt, fell victim to the flu. Doctors could no longer jump on horseback or catch a horse-drawn omnibus to reach the bedsides of their patients, and the epidemic took off, and nothing could stop it. Through a domino effect, trains remained stuck in stations because their necessary fuel, coal, could not be delivered—nor, without the help of horses, could it be extracted from mines, and a well-oiled economic machine came to a grinding halt. The effects of a lack of horses extended to fire pump trucks, which without their "motors" could no longer leave their stations. As a result, in November 1872 Boston, in one of the most devastating fires of its history, saw close to 800 buildings go up in flames in just a few days.

The Horse Today

From those bygone days, there remain large double carriage entrance doors, equestrian statues of our triumphant emperors, and national stables—those lofty places where the reproductive life and breeding of horses are so carefully controlled. There are the legendary names of fabulous horses that have marked our imaginations and collective memory, such as Pegasus, the winged horse of the Greeks, son of Poseidon and bearer of lightning and thunder, or the stormy Bucephalus, who, if we are to believe ancient history, only Alexander the Great was able to tame. Even now, there are 600 or so horse breeds that are fashioned according to our needs, to perform tasks for us, from large draft horses to racehorses, or as companions of the Republican Guard. But there are currently between 7 and 10 million horses in North America and only 60 million throughout the world, a number that is less than the total human population of France.

For more than 4,000 years, the horse has earned the reputation of being the most indispensable and most noble of all human conquests. And yet today it is no longer that inestimable companion outside of a few rare places in the world such as Mongolia, for example, a country with a population of 4 million that has more horses than inhabitants. In the West, the horse has become almost exclusively an instrument of the leisure and sporting industries, which at times use cloning and ever more elaborate doping strategies, borrowing from the most sophisticated genetic technologies, in the pursuit of optimal performances and increased returns on their investment.

Throughout its history, the horse thus tells us a great deal about what our societies once were, but also about their contemporary aspirations. It is that history we will travel through together, borrowing here and there the tools of historians and archeologists but above all those of geneticists. As surprising as this might appear, today we can read the DNA of horses like an open book, including that of horses that now exist only as skeletons in the collections of natural history museums or at archeological digs. We will go back in time, traveling through genealogies and the course of history no longer on horseback, but by way of the messages inscribed in horses' genes. We will discover the great stages of the horse's biological evolution and the changes introduced by breeders since the dawn of its domestication, including the stages that accompanied the creation of the animal we know today. It is precisely that voyage that my research team and I have been working on for more than a dozen years now, without for a moment anticipating the surprises we would encounter along the way. If that voyage has taken us to the four corners of the world and into the depths of time, it has above all led us to rewrite entire chapters of the history of our most noble conquest. Let's now set off without further delay.

2

The Origin Horse

Phar Lap, the Star of the Interwar Period

"Hi Ludovic, I'd like to introduce myself. I am the filmmaker who requested the DNA authentication of Phar Lap's heart specimen." I must confess that up to then I had never heard of the champion racehorse until the curators overseeing his remains had contacted me a few months earlier. When Kerry Negara wrote that email to me, I was not entirely aware of just how legendary that English Thoroughbred of the 1930s remained on the other side of the planet, in Australia. So legendary, in fact, that people like Negara were still producing documentaries or investigative series for the radio on his life. Negara indicated that her family had authentic hair taken from Phar Lap's tail, directly provided by their very close friends, Tommy Woodcock, Phar Lap's trainer, and his wife Emma. She offered to come meet me and give them to me personally, if these could help to authenticate Phar Lap's genetic profile. The project intrigued me; I would have to learn more. The ABC News piece I immediately fell upon would pique my curiosity even more. Its title: "Expert Confirms Phar Lap Arsenic Theory."

This animal's story is fascinating, and worthy of the best Hollywood scripts. The champion was born in 1926 in New Zealand, where he was given his name, which in Thai means "lightning." It turned out to be the perfect choice, because despite rather sluggish initial competitions, the animal quickly proved worthy of its name. He won the most prestigious races for three-year-olds, as well as the mythic Melbourne Cup, the most coveted crown of all the speed races in the Southern Hemisphere. There was then nothing but an almost uninterrupted series of wins in Australia, until, at the end of 1931, the animal traveled to the United States, accompanied by his glorious reputation as an invincible racehorse.

Beyond the promise of marking history with new racing exploits on the other side of the Pacific, the prospect of the particularly lucrative associated prizes must have played a role in the decision of the horse's owner, a businessman, to have the animal live the American Dream. Phar Lap made a first stop in Mexico, where on March 20, 1932, he won the Agua Caliente Handicap, the highest-paying race at the time, crushing his competition: he crossed the finish line two lengths ahead of the other horses, and in doing so won a $50,000 prize. When Phar Lap later went to a private ranch in Menlo Park, California, to prepare for his coming competitions, the world of horse racing was trembling in anticipation. It is reported that King George V sent a telegram to celebrate "with all his heart the wonderful victory" that Phar Lap had just enjoyed at Agua Caliente. There was even a plan to meet with Metro Goldwyn Mayer to discuss the possibility of producing a series of short films about him. The animal was at the summit of his glory. No one could have imagined that he would never again walk onto a racetrack—and that on the following April 5, in the early morning, Tommy

Woodcock wept over the body of his soulmate, who had died after a night in terrible agony.

A Scandalously Suspicious Death

The circumstances of Phar Lap's death caused much ink to flow. Accident, overdose, infection, and even poisoning—just about everything was imagined. Admittedly, following the financial crisis of 1929, in the midst of Prohibition, some people were ready to do anything for money. It must also be said that a champion of Phar Lap's caliber could easily have upset the business model of organized crime and its armies of bookmakers. The *Chicago Daily News* estimated that between the activities connected to illicit bars and those linked to bookmaking, $6 million and maybe more must have filled the coffers of the man who at the time was the true godfather of the city: Al Capone. When in 2010 a German team discovered that Phar Lap's hair contained massive quantities of arsenic in a chemical form that was different from the kind used in taxidermy, one of the most colorful theories explaining the animal's death again focused on the animal's hair: Phar Lap had been poisoned because he was too great a threat to the financial interests of the crime boss himself.

The theory was credible; at least that's what Negara thought, even if the true perpetrator and motive for the crime would surely remain forever a mystery. This is why she wanted to have genetic tests done on Phar Lap's heart, which is still on display in Canberra, as an icon in the collections of the national museum. Her own theory was that the heart would prove not to be that of Phar Lap, but of another horse that had been sacrificed to mask the crime and prevent laboratories of the time from detecting the presence of arsenic in the tissue. The death would then be classified as natural, and the affair would be filed away without any further investigation that might threaten the true criminals. It's true that the laboratories of the 1930s were not equipped with a synchrotron, which had enabled the German team to discover the spike of arsenic in the hair they analyzed. Poisoning as the cause of death could hardly be determined unless tissue such as that of the heart were analyzed, thus the absolute necessity for subterfuge. In Negara's opinion, we would just need to obtain Phar Lap's genetic profile using the hair she had and compare it to that of the heart to reveal the ruse. And turning to my lab, we would then be able to extract DNA from those vestiges to test her theory.

An Inspiring Failure

Unfortunately, the heart had been soaking for close to a century in chemical products such as formaldehyde, which embalms tissue and prevents its post-mortem decay, but for that very reason often makes geneticists' most sensitive tools ineffective. Under those conditions, it was impossible to take the analysis as far as Negara had hoped. We did, however, obtain DNA from the bones of other champions that lived in the early twentieth century: Dark Ronald, another English Thoroughbred born in 1905, around twenty years before Phar Lap, and the Baron O'Buchlyvie, known as the most expensive draft horse ever sold at public auctions, for the equivalent of $1.6 million today. It was in 1911, three years before the world record holder from Glasgow died at the age of fourteen.

The genetic profiles we obtained for the two animals were consistent with photographs; they did not carry gene versions deriving a chestnut coat or lighter bay coat colors, for example. But our analyses also revealed things that photographs can hardly tell: both animals carried DNA associated with anxiety and agonistic behavior. They also turned out to be less inbred and the breeding stocks that produced them were larger and more diverse than in modern Thoroughbreds or Clydesdales. For lack of analyzable elements we hadn't succeeded in corroborating Negara's theory, but we had nevertheless deciphered part of the genetic information from other legendary animals—and in doing so, shed new light not just on their own biological characteristics, but on the recent history of breeding.

If we were able to use genes to confirm, close to a century after their death, that Dark Ronald was indeed an English Thoroughbred and the Baron O'Buchlyvie a Clydesdale, and see characteristics that are not measurable in skeletal bones, couldn't we then hope for similar success on their predecessors? By this I mean all the horses who had lived before them, whether those famous enough to still be preserved in our museums of natural history, or the legions of unknowns who rest in archeological sites. And if this proved possible, couldn't we then go back in time and use DNA to rewrite the history of the domestication of the most noble of our conquests, and thereby reconstruct a genetic history of the horse?

Once it had taken shape in my head, I became increasingly obsessed with the idea—but I didn't yet know that I was going to devote more than ten years of my life to it. Today, a decade later, I am happy to be able to confirm that yes, we can take that voyage back in time; yes, we can travel back in history and read DNA the way others read ancient texts; and yes, we can thereby solve some of the greatest mysteries in archeology. In particular, these: Which of our ancestors were the first to have domesticated the horse? Where and when did they live? And how did they succeed in domesticating one of the last of the large wild mammals?

Tracking the First Domesticated Horses

"The Earliest Horse Harnessing and Milking." This is the title of an article published in the scientific journal *Science* at the beginning of 2009, whose lead author was my colleague Alan Outram, a professor at the University of Exeter. It seemed like a very good point of departure. In the paper, Outram discussed the discoveries he had made in Botai, in Kazakhstan, at an archeological dig located a few hours' drive northwest of Astana, the city whose name means "the capital" and which in 2019 was renamed Nur-Sultan in honor of the president Nursultan Nazarbayev, when he handed over power after having led the country with an iron fist after the fall of the Soviet Empire. The Botai site is so important that the name was subsequently used to designate an entire culture—the Botai culture—vestiges of which have been found at other nearby sites, such as that of Krasny Yar. It's important to point out that this culture, though located right in the middle of the central Asian steppes, was one of a sedentary, not nomadic, people. People lived in round huts, half-sunken in the ground, a type of yurt not made of felt and fabric as in Mongolia, but of tree branches and clay. The dwellings were built to stay in place. It was impossible to miss them on the terrain in the summer, even for a lab rat such as myself; the grass there grows as in small yards, causing patches of green placed next to or opposite each other to appear over the steppe, thereby revealing the places where the people once lived. Once the quadrat frames were placed and the dig underway, the floors of the houses soon appeared under scarcely a few dozen centimeters of sediment accumulated over 5,100 to 5,500 years. Fragments of clay pots, on which triangular geometrical motifs, filled with parallel dotted lines are drawn, were also sometimes found.

Botai: The Cradle of Domestic Horses?

Shallow ditches are scattered over the terrain between the habitations, sometimes even inside them. One can find large accumulations of animal bones in them. Viktor Zaibert, who discovered the site in the 1990s and excavated it almost every summer until his recent death, was not a little boastful when he reminded me that more than 300,000 animal remains were discovered in Botai. But what distinguishes the Botai culture from other cultures of the time is that just one animal species dominates in the deposits. The species is neither the cow nor sheep, as are found among most pastoralist peoples, nor the dog, but the horse. Almost all of the remains recovered were only those of horses. Botai people were thus horse people. However astounding these figures may appear, they do not necessarily imply that breeding took place there or that horses were domesticated; they might have been hunted (granted in very large numbers), but hunted all the same, and not bred.

However, ethnography teaches us that hunter-gatherers were rarely completely sedentary, since their subsistence depended on their ability to follow their prey. Years later, our genetic work would show that populations of horses were in demographic decline in that same period. It is difficult to imagine a people primarily hunting a resource that was becoming increasingly sparse, while also remaining sedentary and not leaving to explore other territories where their prey might be more abundant. Other elements in Outram's and Zaibert's discoveries also seem to point to domestication. The second premolars of some Botai horses show traces of wear, which according to specialists was not from normal use and could only have occurred through repeated contact with a hard object. However, digs in Botai have not uncovered any sort of metallic bit, nor bits carved from bones or from the antlers of reindeer; these wouldn't appear in great numbers until more than 1,000 years later. Ropes or straps must have been used to restrain horses, but since such objects degrade quickly, in just a few years, they only survive under exceptional conditions in archeological vestiges over 5,000 years old. Lacking irrefutable physical elements, archeologists are reduced to adopting that interpretation for explaining the wear of the teeth. If the interpretation stopped there it would be far from convincing, and other researchers have contested it. But there is something else.

The hunter's work doesn't stop when the prey is killed; once it is dead, it still has to be cut up and carried back home. A horse weighs around a half-ton. It would be difficult even for a group of hunters to carry the entire animal back to their dwellings without a huge effort. So hunters selected the most prized portions of the animal—those that contained the most muscle, fat, and bone marrow. This is a well-known fact in ethnographic studies and has been verified by studying bones found at sites dating from periods when the animal had not yet been domesticated (most likely hunting sites). If the Botai people lived solely from hunting, the multitude of horse bones found in ditches at their dwelling site should predominately represent the parts of the carcass they preferred. One would then find signs of what specialists refer to as the *schlepp effect*—once butchered, parts of horses are discarded at the kill site, while other meatier parts are taken back to the dwellings. But if horses were domesticated, killed, and eaten on-site, one would expect to find, with a few exceptions, all the parts of their bodies in appropriate anatomical proportions. This is precisely what Outram's and Zaibert's observations seemed to indicate: there was no trace of the schlepp effect in Botai; however, there were signs of wounds to the head suggesting a violent death, a type of death blow that would have been administered to a restrained, immobilized animal, one felled by a single blow of an axe—not the type of wound of an animal being pursued. The image of a pastoralist people breeding horses was becoming clearer. But there is even more.

Moreover, the disposition of the huts in Botai, as well as in Krasny Yar does not appear random. One can easily make out an alignment of huts delimited by paths, sometimes even what would look like squares. The Botai village had been the result of an urbanization plan, so to speak. The essential point for us is that archeologists have also found deeper holes there into which poles were placed to hold fences, delimiting a space (often hexagonal) where animals were corralled. We know this because the ground inside the enclosures is rich in organic compounds containing nitrogen and phosphates characteristic of animal dung. These were enclosures where horses were held and kept, near the habitations; everything suggests that they occupied a central place in the lives of those people. It seems the horse was domesticated, in the sense that it was bred on-site. At this stage we begin to understand why Alan Outram, Viktor Zaibert, and many other archeologists consider Botai and its environs the cradle of the domestication of the horse.

Yes, but . . .

Some archeologists resist this argument and remind us that the size of the Botai horses wasn't different from that of the horses that lived in the wild in that region or elsewhere. A reduction in size for cows, for example, as compared to its wild ancestor, the aurochs, is noticeable starting at the very beginning of its domestication; size is one of the primary indications that unearthed remains belonged to animals that were no longer completely wild and had been domesticated. By this logic, the Botai horses therefore could not, in their opinion, be domestic horses. But for horses the first changes in size appeared only later in history; they go back to the Iron Age, more than 1,000 years after archeologists discovered entire chariots harnessed to horses. It is obvious that to be harnessed horses need to be domesticated in the first place, even though their size was in no way a distinguishing feature. Indeed, who would take the risk of trying to harness a wild animal to any sort of vehicle and allow him- or herself to be led by a beast that couldn't be controlled? The size argument was far less dismissive than it first seemed to be in disqualifying the interpretation of Outram and Zaibert.

Other archeologists have voiced doubts that there was husbandry in Botai, insofar as breeders often prefer not to keep a large number of males in their herds. Generally speaking, only a few genitors suffice to ensure the next generation, and breeders keep a larger number of females in the hope of producing more animals. However, in Botai, the remains of male and female horses have been found in fairly equal proportions. The proportion of remains identified as foals doesn't seem particularly higher either, compared to that of adults. If the Botai breeders didn't sacrifice their youngest males, could they really have been breeders? Wouldn't they simply be hunters? Doubt continued. However, if the Botai only hunted, one would again expect archeologists to discover a larger number of foal bones, mainly because young horses as prey would be easier to catch, but also because horses live naturally in social groups protected by one or several males and in which the females band together with their foals. An attack on a group, therefore, would certainly lead one to expect large proportions of juveniles. Here again the argument advanced against Outram and Zaibert's theory did not advance the debate. What is more, the breeders might have wanted to keep their adult male horses for other reasons, such as for transport or for hunting. After all, which animal other than a horse could enable a hunter to run down and catch his prey? According to this interpretation, the horse would have been domesticated in Botai to facilitate access to resources that were becoming increasingly rare. Husbandry guaranteed on-site access to meat, and riding would have possibly enabled food resources to be supplemented through hunting on horseback. Outram and Zaibert were convinced of this, but skeptics remained.

Elemental Traces of Horse Milk

Then a new observation was made. This time it didn't involve the horses' teeth or bones, but the clay pots used by the people of Botai. Outram and Zaibert had noticed that food residue had sometimes been preserved on these pots, residue that seemed to have stuck onto the surface and had endured in the ground for more than five millennia. Could a chemical analysis be done to determine its nature? This is when Outram had the brilliant idea to combine forces with Richard Evershed, a professor at the University of Bristol, not far from Exeter. Evershed is a geo-chemistry expert; he knows how to analyze elemental traces present in archeological vestiges, specifically those that are known as isotopes. Without diving too deeply into physics concepts, let's just recall that the nucleus of an atom consists of positively charged protons and electrically neutral neutrons. Other electrically negative particles form a cloud around the nucleus, whose volume is considerably greater than that of the nucleus itself. If a given atom—hydrogen, for example—always has the same number of protons (a single one in this case), it will not always have the same number of neutrons. If it doesn't have any, it is ordinary hydrogen; if it has one, it is deuterium, and if it has two, it is tritium. Ordinary hydrogen, deuterium, and tritium are what we call isotopes, that is, variants of the hydrogen atom that differ depending on their number of neutrons. The same is true for carbon; most of the time we find it with six protons and six neutrons, but sometimes, too, in its nonradiogenic forms, with one additional neutron. It is then called carbon-13 $(6+7)$ instead of carbon-12 $(6+6)$. What is important for us is that an atom's mass will vary depending on the number of neutrons it carries, and that difference, as small as it may seem, means that it will not behave entirely the same way in nature.

This difference will be especially reflected in the molecules that make up living organisms, because those molecules are formed following numerous and sometimes complex biochemical reactions that, depending on the isotopes involved, do not always occur at the same speed. They are slower for isotopes weighed down by their supplementary neutrons, such as carbon-13 as compared to carbon-12. The difference may be infinitesimal, but it has observable effects if we consider all the reactions that occur in us; our bodies end up being globally deficient in carbon-13 compared to the level at which the isotope is naturally present in the environment. Depending on the number and type of reactions involved in its formation, a certain biological tissue can be more or less deficient than another, and thereby acquire an isotopic signature—an isotopic value, if you prefer—that is unique to it. This is true, for example, of the fatty acids that make up adipose tissue, or of those that one finds in milk; although they are both comprised of fatty acids, each contains isotopes in specific relative proportions, and therefore presents a different isotopic signature that can be used to distinguish them. In that case, couldn't we use these same signatures to determine if the people of Botai consumed horse milk, or if they instead consumed their fat?

Evershed had described the principle ten years earlier, which is why Outram turned to him. There is fat and then there is fat: fatty acids are diverse and contain characteristic proportions of different isotopes of carbon atoms, which leads to measurable differences in mass. Palmitic acid, for example, contains a chain of carbon atoms that is longer than that of stearic acid, and their isotopic signatures are therefore not the same. Because the composition of fatty acids in the fat of different animals is not exactly the same, repeating the analysis for each type of fat can tell us which animal we are looking at: horse, swine, or ruminant. In theory, we could therefore not only know if it is milk or muscle fat that was consumed, but also from which type of animal it came.

Scientists just had to extract the fatty acids encrusted in the pottery shards using chemical solvents such as hexane, methanol, or chloroform before volatilizing them in the form of gas to be able to analyze them in a mass spectrometer. This machine is built specifically to quantify even the most subtle differences in mass, and can show us which isotopes predominate in the residue left from the meals of the people of Botai. Clearly, and not surprisingly, the isotopic signatures found in palmitic and stearic acids confirmed the consumption of horse. However, although the analysis easily distinguished milk from pure fat in ruminants, the same analysis proved difficult in discriminating them in horses. Another isotopic signature had to be studied in order to make progress.

Evershed suggested looking at another isotope, deuterium, because in central Asia the isotopic composition of water (which contains two hydrogen atoms) is not the same when water falls in the form of snow in winter as when it falls in the form of rain in summer. Seasons, just like the bodies of animals, also fractionate isotopes, and particularly in continental climates with very low temperatures in the winter and very high temperatures in the summer. A mare doesn't produce milk in the winter, only in the summer; the isotopic composition of the milk will reflect that of summer water and will be enriched in deuterium. The same is not true for fats, which are renewed both in the summer and the winter and present somewhat equal amounts of isotopes. The experiment was conducted, and the fatty acids encrusted in the clay shards proved to be quite specifically enriched in deuterium. The residue was from the milk of mares and not from fat. It seemed that there was no more room for doubt, even for the most recalcitrant skeptics: the people of Botai were a pastoralist people, involved in breeding horses. That husbandry served as much for procuring milk as for procuring meat. Breeders must also have used their horses for transportation, maybe even to capture or hunt other horses more efficiently than on foot. In other words, their culture was an all-encompassing horse culture.

The DNA of the Botai Horses

It was to Botai that one had to look if one wished to genetically describe the very first domesticated horses. So I contacted Alan Outram, whom I didn't know at the time, in the hope of gaining access to the horse bones to carry out their genetic analysis: ideally to reconstruct the entirety of their genetic information, their entire genome, a sequence close to 3 billion letters long. The technologies had come a long way and my team had just successfully carried out a sequencing not on a horse only a century old (such as the Baron O'Buchlyvie, who had died a month before the First World War started), but on an organism more than a half-million years old, between around 560,000 and 780,000 years old—a horse, still.

By employing the same technologies on the horses of Botai, we could hope to better categorize those horses and discover things that are impossible to see in just teeth and bones: the color of their coats, for example, or the fact that they were or were not carriers of genetic diseases common in some horses today, or to determine for which physical, physiological, or behavioral traits their breeders had selected them. Outram was enthusiastic and immediately agreed to collaborate. I soon received in the mail an initial set of samples taken from the Botai horses, which we were eager to analyze. I was completely unprepared for what we would discover. I went to Botai in the summer of 2016 to collect new samples in person. Our surprise at our results was such that we successfully repeated the analyses to confirm their validity.

Our surprise was infectious, so much so that the journal *Science* decided to devote a full cover, illustrated with a splendid photo of a horse, to our discovery. However, the illustration didn't depict a Thoroughbred horse, a racehorse specifically representing the most noble standards of domesticated beauty. Instead, the journal chose a Przewalski's horse—otherwise known as the wild Mongolian horse—to represent the closest relative of the horses of Botai. No image could have better captured the very essence of what we had just discovered: the horses of Botai were not the ancestors of our domestic horses living on the Earth today, whether working fields, running on racetracks, or serving for leisure equestrianism. Instead they are the direct ancestors of Przewalski's horses, the very ones we had always assumed, since their discovery around the end of the nineteenth century, to be the last wild horses living on the planet. We had to confront the evidence: the Przewalski's horse is not exactly the wild being we had imagined, but it can be thought of as the descendant of the first wild horses ever to be domesticated — the descendant of the horses of Botai, simply returned to a wild state. It seemed incredible.

And our discoveries didn't stop there, because in addition to the genomes of some twenty Botai horses, we were also able to sequence the genomes of around forty other ancient horses. Some were also from Kazakhstan, such as those from the site of Borly 4, in the Pavlodar region, which radiocarbon dating proved to be 5,000 years old; or those of Grigorevka 4, located nearby but more recent at 1,200 years old. Others came from the Gallo-Roman era in France or from the Carpathians and dated from the Bronze Age. There was a whole collection of specimens encompassing vast areas between the Atlantic coast and Siberia, passing through Byzantine Anatolia, Achaemenid Persia, the Xiongnu region of Mongolia, and many other locations. By sequencing their genomes, we hoped to understand how horses had left their original cradle in Botai and its environs and expanded to the rest of Eurasia. In short, we expected to recount the genetic history of the horse and understand its conquest of the world. We could not have been more wrong. Not one of the horses of those great civilizations of the past was related to those of Botai; not one of them had descended even remotely from the Botai horses. This discovery had a major implication: the domestication of the horse had not been a unique episode in the history of that animal. In other words, there hadn't been a single and sole domestication in Botai, and clearly the horse had been domesticated elsewhere as well. We just didn't yet know where that might be.

Breaking News: Domestication Beyond Botai

How did DNA enable us to arrive at this conclusion? First, let's note that we initially sequenced each of the chromosomes of each ancient horse in their entirety: the thirty-one chromosomes that are inherited from both parents (called autosomes), sex chromosomes (X and Y chromosomes, the latter being carried uniquely by males), and the mitochondrial chromosome, a total of several billion combined letters from a genetic alphabet consisting only of four (C, A, G, and T). In each generation there appear random variations in that sequence, because the divisions that lead to the formation of reproductive cells, sperm and ova, do not always copy identical genetic information but instead introduce changes—mutations—here and there. Granted, a newborn foal will have inherited one set of chromosomes from its mother and a second set from its father, but the sequences that come respectively from each of the parents will marginally include a handful of variations.

On the scale of a population, at the same place on the genome, the genetic sequence can exist in several versions with different letters and constitute what population geneticists call a polymorphism. Most often, the different versions of the sequence will have at most a marginal effect on biological functions. This is called neutral polymorphism: mutations appear randomly, the total sequence is long, and the genes represent only a limited portion of it. On average, neither the horses' life expectancy nor fertility are affected by it. Under these conditions, generation after generation, the transmission of one or the other of the versions of the sequence depend only on chance and the number of mares and stallions that are crossbred. Let's briefly see why.

Imagine the extreme case of a population reduced to a single couple: on the one hand, a mare carrying at a given place on her genome the same letter twice, let's say A; and on the other hand, a stallion carrying at the same place the letter A and another letter, let's say G, due to a mutation. The mare could only transmit to each of their offspring the single letter she carries, whereas the stallion would transmit half the time either A or G. It's like a game of heads or tails. In this game, if the couple is very prolific and brings ten foals into the world, there is a good chance that polymorphism will be maintained in that generation: each of the two sides of the coin, A and G, will have likely been obtained at least once out of the ten tosses; the probability of obtaining only "heads" ten times in a row is negligible (less than one chance out of a thousand). By contrast, if they only have two offspring, it is possible that both tosses will result in two "heads," A for example (one chance out of four). In that case, the polymorphism present in the parental generation will have disappeared in the next. We can thus see that what determines how long a neutral polymorphism will remain in a population—that is, the time between the appearance of a mutation and its disappearance (no one carries it anymore), or its fixation (everyone carries it)—is above all determined by the demography, or the number of individuals participating in reproduction. Geneticists call this the effective size of the population. The larger the effective size, the longer a polymorphism will remain present, and the longer the time two versions of a sequence will coexist in the population concerned. In contrast, when the effective size is smaller, the frequency variations of polymorphisms from one generation to the next (this is called genetic drift) will be more noticeable and the amount of time a polymorphism will occur is shorter. One arrives more quickly at a situation in which only one version of the sequence will subsist at the relevant place in the genome.

What we should remember is that once populations diverge and cease crossbreeding, the differences in the genomes of the individuals that make up each population accumulate independently, because they appear randomly. The separation in relation to the initial sequence will reflect the time passed, since the divergence and populations of reduced effective size will see great variations in the frequency of the polymorphisms they contain. Such situations can occur when an ancestral population initially occupying a given geographic zone splits in two, such as when a handful of individuals manage to reach the coasts of a new island to settle there, or when humans separate one group of animals from others by domesticating it. The differences found in the genomes of individuals belonging to two different populations offer us an indication of the time elapsed since they diverged—it offers us temporal information. Furthermore, the differences discovered in pairs of individuals of the same population give us an idea of the effective size of the population since its formation: we obtain demographic information, which we can see as the average number of individuals having reproduced over generations.

The genomes we sequenced in horses enabled us, through simple comparisons, to say which ones were members of the same population (if the genetic sequences were very close) and which were members of different populations, and in that case, to determine the amount of time those populations had been separated, and even to have access to their respective demographic trajectories. This comparison of sequences enabled us to reconstruct what population geneticists refer to as their evolutive history. As we can see, the genetic sequence can be read and understood as a new historical archive—even for species other than ours that have never developed writing.

Going back to our study, this comparison of sequences revealed the existence of a significant separation between Przewalski's horses and domestic horses living today. The divergence between these two branches would have begun, according to our estimates, around 35,000 years ago. As for the Botai horses, they appeared on the first branch, the one that led to Przewalski's horses: they were thus their direct ancestors. Apart from the horses of another Kazakh site (Borly 4, which is 5,000 years old), none of the other horses we sequenced appeared on that branch. Everything seemed to suggest that the Botai horses' direct descendants were those of Borly 4 before that lineage disappeared, reappearing in the form of Przewalski's horses. Obviously the lineage didn't completely disappear, since Przewalski's horses are still around. The apparent disappearance was the result of our sampling, which was clearly linked to the archeological sites where humans lived close to their horses. Our results suggested a history of distancing from humans and of a return to a wild state, rather than that of a disappearance. Furthermore, the reconstructed evolutive history noted a particularly strong genetic drift in that interval. The demographic history of the Botai horse was therefore one of a notable reduction, not an expansion, unlike that describing the irresistible spread of domestic horses. Moreover, in that initial work we were able to discover the first genetic trace of the latter in a horse that lived around 4,000 years ago in the Carpathian Basin, in Czechia. All the other horses we sequenced were grafted after it along the same genetic branch, which was distinct from that leading to Botai and Przewalski's horses, regardless of their age and their geographic origin. We were able to conclude that somewhere between 5,000 and 4,000 years ago, during the third millennium BCE, the first domestic lineage of Botai had been replaced by a second domestic lineage, the one that led to our modern domestic horses.

It was indeed a replacement, because the comparison of genetic sequences indicated hardly any crossbreeding between the branch to which the Botai horses belonged and the second lineage, which for the lack of better inspiration we called DOM2. In this second lineage we found only around 3 percent of the genetic variation present in the former; in other words, looking only at Botai, 97 percent of the genetic variation characteristic of DOM2 horses remained unexplained. The origin of modern domestic horses remained unknown. To find it, we had to look elsewhere, beyond Botai. We knew which period in the past we had to explore—the third millennium—but that's all we knew. Who were the people who had domesticated the horse the second time? Where and when did they live? How were they able to tame one of the last large wild mammals left to be domesticated? We had almost returned to our point of departure.

3

The Other Origin of the Horse

The Tarpan of the Western Russian and Ukrainian Steppe

It was July 23, 2018. My plane landed late at night at the Samara airport where Pavel Kuznetsov, a professor of archeology, was waiting for me. He is a connoisseur of this region, the western steppe of Russia, and a specialist of the domestication of horses. Kuznetsov is one of the scientists who never adhered to the idea that the Botai horses were the ancestors of our domestic horses today. One might even say that he is one of the most ardent detractors of Alan Outram's theses. The results we had just published at the beginning of April, which revealed no genetic affinities between Botai horses and our modern domesticates, were right up his alley. In his opinion, it was in this region in the world, and not in central Asia, that we needed to seek the authentic cradle of domestic horses.

He would tell me this a few days later when he took me to visit several major archeological sites in the district of Krasnoyarsk, one of the twenty-seven provinces under the administrative umbrella of the Samara Oblast. I can still see him turning toward me and pointing his finger in the distance, toward the southwest, before adding a few sentences of which I managed to understand only a few snippets: "Ludovic," my first name, followed here and there by the rare Russian words I know, "ЛОШаДЬ" and "ТарПаН." To my francophone ears, it sounded like "loshet" and "tarpaniye." What those words referred to, however, was enough to immediately catch all my attention: "horse" and "tarpan." Fortunately for me, the zooarcheologist Natalia Roslyakova was with us and served as interpreter; her summaries were particularly useful to me. What Pavel just told us, she explained, was that at hundreds of kilometers from here, in the expanse of the steppe that leads to the Caucasus, southwest of where we were at the time, there lived a wild horse called the tarpan. It was this horse that the ancestors of the people who live there had one day captured and were able to domesticate—and therefore *the* horse, and no other, according to Pavel, that gave birth to the modern horse. In a few simple sentences he had just summed up the other possible origin of the horse, the one that we would subsequently explore.

Earlier, when my plane had just landed, Kuznetsov had come alone to pick me up at the airport, and our conversation was necessarily rudimentary for the entire journey to my hotel. Happily, the language barrier would soon no longer be a problem, because in scarcely a few days spent together moving hundreds of boxes, crates, and bags all filled to the brim with archeological vestiges; picking out the remains of horses from among those of other animals, including humans, as well as bits of ceramics; and separating out tiny fragments of bone or teeth, a true bond was formed between us. So much so that I brought back home his family recipe for bortsch. For the specimens that I had retrieved, I still had to wait for all the administrative authorizations to be duly obtained—even though Kuznetsov was the head of the collections of the archeology laboratory, and even though the rector of the Samara State University, whom I had met during my visit, had given us his blessing. I was confident that these remains would make it back to France and to my lab, the Centre for Anthropobiology and Genomics of Toulouse. Several years later, some of them, like those from the Turganik sites, would prove to be essential pieces of the puzzle.

The Beginnings of a New Relationship

Kuznetsov wasn't the only one to think that the western steppe of Russia could have played a central role in the domestication of the horse. David Anthony, with whom Kuznetsov had done a lot of work, was of the same opinion. Although Anthony lived on the other side of the Atlantic, working until his retirement as a professor of anthropology at Hartwick College in New York, he spent a large part of his career studying the archeology of Samara and the surrounding area. He is among those who recognize, in sites located a stone's throw from the town, the earliest signs of a profound change in our relationship with the horse—as at the Khvalynsk cemetery, for example, on the left bank of the Volga River, known since the end of the 1970s. It is there that the graves of nearly 160 people who died around 6,500 years ago are found, as well as the remains of a great many animals, ostensibly placed as sacrificial offerings to celebrate the memory of the dead and accompany them in their final voyage. They are ruminants for the most part, but one also finds a not-negligible number of horses—not their entire bodies, rather only their legs. A deliberate choice, then, suggesting the presence of a funerary ritual. Among the objects found alongside the remains are also weapons such as hammers, whose polished stone heads easily recall animal heads, including horses.

Khvalynsk is not an isolated case, because a very similar scenario seems to have played out in S'yezzhe, a burial site near Khvalynsk known for revealing the vestiges of another horse, also apparently sacrificed. This time the people had chosen to place in the grave only the animal's hooves and skull, and not far away were two caballomorphous figurines that to all appearances looked like horses. For our anthropologist, these sacrifices suggested that the nature of humans' relationship to that animal was in the process of changing, and that the horse was becoming a central element in the beliefs of peoples who had preceded the Botai by a good millennium. At that time, the horse was perhaps still only prey for the hunt. Nevertheless, its status and its place in our societies were no longer completely the same: it had clearly left the simple sphere of being a source of food and had assumed a more symbolic, possibly even religious dimension, as it was associated with funerary rituals. Perhaps this was enough to propose the hypothesis that it had already been domesticated, and that the process later leading it to replace the horses of Botai during the third millennium BCE had already begun. This is in any event a hypothesis that shouldn't be rejected out of hand, if we are to believe David Anthony, especially since other sites, also older than Botai, seemed to point in that direction.

One of those sites wasn't located in the Russian steppe, but in that of neighboring Ukraine. Its name was Dereivka, and it was gigantic, with close to 3,000 square meters of habitations. Without equaling the numbers found in Botai, the proportion of horse bone remains found within this vast expanse was close to two-thirds as much—enough to provide more than 7 tons of meat, according to specialists. The horse therefore had an important place in the food supply of the people who once lived in this territory. However, what was most spectacular was not the monumental nature of the whole, but the vestiges of just one of the animals that was preserved there. The head and left foreleg of a stallion, who, judging from its teeth, must have been seven or eight years old at the time of its death, had been placed with the complete bodies of two dogs next to it. Not far from them was a clay figurine in the form of a pig and round objects carved out of reindeer antlers, apparently the elements of a bridle. Here again, the horse figured at the center of a funerary ritual. But that wasn't all: its teeth, in particular its second premolars, were extremely worn down, as if their front facet had been planed. Only repeated grinding on a bit could have created such beveling, that much was clear. The stallion was clearly not only the central element of a funerary ritual or a religious ceremony; it had also been ridden its entire life. It seemed that it wasn't just on the banks of the Volga, but also on those of the Dnieper, that people had begun to domesticate the horse and travel on its back for the first time, around 6,000 years ago.

That was at least one of the models that prevailed to explain the origin of our domestic horses, until David Anthony subjected the primary Dereivka specimen to radiocarbon dating, aiming to situate this key element in the history of our civilizations more precisely in time. A new dramatic turn of events: far from being 6,000 years old as expected, the stallion appeared to have lived only between the eighth and the third centuries BCE. It came directly from the Iron Age and corresponded to an intrusion of more recent archeological strata that hadn't been noticed during the dig. This specimen didn't support the proposed theory, which was then weakened. Coming from a period when mounted horsemen had already existed for a long time, it told us nothing about the beginnings of domestication. Looking more closely at the dental remains of the other horses—the ones that weren't intrusive—one could deduce their age at death, which indicated a peak mortality of between five and eight years old. That is the very age when animals are in their prime, which defied all logic of breeding; that logic should aim to cull most of the males when they are very young, and females when they become too weak to be expected to continue giving birth. It seemed that the death knell of Dereivka, as the steppe cradle of the domestication of horses, had sounded.

Looking for the Needle in the Haystack: The Hypothesis of a Home in Central Europe

But did the hypothesis of an origin in the Russo-Ukrainian steppe have to be forever abandoned? Not necessarily, but we couldn't allow ourselves to reject other regions of the world a bit too quickly, such as central Europe, notably Hungary or the Czech Republic, where the fossil record indicates that the size of horses strangely changed during the sixth and third millennium BCE. At that time not only did the average size increase, but it also became more variable from one horse to another, before evening out on the low end beginning in the middle of the third millennium. The fact that adult horses of the same population began to reach sizes that were sometimes above and sometimes below those they reached naturally in the past would be precisely the sign that the horse was no longer evolving in nature by itself, but was undergoing the effects of human intervention. Either humans had improved feeding for some horses, thus encouraging their growth, or they had protected the smallest among them, thereby fostering their survival. That their size later ended up being standardized and reduced would ultimately be the sign of a more advanced control of breeding: either for practical reasons, because it is easier to work with animals that are all similar and one takes fewer risks if they are smaller; or for economic reasons, because it is possible to fit more animals into corrals if they remain of modest size.

The argument may seem convincing at first, but it nevertheless remains light years from being proven. There are numerous parameters that intervene to regulate the size of an organism. There is the temperature of the environment in which it lives, for example: Bergmann's rule reminds us that the size of warm-blooded animals tends to increase near the poles and to decrease the closer they approach the equator. Changes in size conveniently interpreted as mirrors of domestication might therefore in reality be only the reflection of slight changes in the climate of the region at that time. Furthermore, subtle changes in the natural ecology of the region, leading to the disappearance of a certain plant or the appearance of another, might also have had a true impact on what horses ate, and therefore on their size. And we mustn't forget that the encounter between two wild populations with obviously different size profiles might very well produce the illusion of a population bred by humans; once the fusion occurred, wouldn't it have shown even more variable size profiles?

Clearly, the hypothesis of a domestication in central Europe, based essentially on an analysis of size profiles, wasn't infallible, to say the least. But it was on the table, and we couldn't ignore it. I contacted René Kyselý and Lubomír Peške, two Czech archeologists enthralled with this type of analysis. Their 2016 paper showed that they possessed very rich archeological material from this region. Perhaps they would agree to have it analyzed by our genetic tools? My email was sent from the Beijing airport, where I had connected to return home from one of my stays in Mongolia. Their response was already waiting for me in my inbox when I landed in Europe: they were ready to help, and would soon send me a wealth of samples.

The Iberian Peninsula, That Other Possible— but Controversial—Cradle

Even if it meant considering all hypotheses, we couldn't ignore the Iberian Peninsula, because that region came up again and again in publication after publication. This land south of the Pyrenees is often presented as the one that offered a refuge to many animal species, including ours, during the Last Glacial Maximum around 20,000 years ago. It is one of the rare regions in the world where humans and horses were able to live together continuously and get to know each other for the longest period of time. Cave art there bears witness to the fact that the people of the region were keen observers of the animal, because it appears as one of the dominant subjects in their artistic representations. What is more, the modeling of ecological zones in the past did not disprove horses living in the region through the ages without ever disappearing. If we inventoried all the archeological remains of horses that have ever been discovered in Eurasia that have undergone radiocarbon dating and place them on a map, we would learn about the climate conditions that were tolerated by horses throughout their history—their climatic niche, in a way. To do this we had only to cross-reference our data with the ever-more precise models produced by paleoclimatologists.

It was precisely that work that we finished in July 2018, on the heels of the genetic study of the Botai horses. The analysis confirmed that throughout the last 40,000 years, the Iberian Peninsula had always offered climatic conditions completely compatible with the ecological demands of horses. Therefore it was perfectly possible that they were able to live there without interruption, and perhaps even that they had ultimately been domesticated there—who knows? The same analysis also revealed another candidate region, in a zone that we are beginning to know well now, beginning on the shores of the Caspian Sea and covering a large part of central Asia and going through Botai. It was proof that these climate models were perhaps not just full of hot air. At the same time, they indicated that most of the other regions, including central Europe and Ukraine, at certain points in their history, seemed to have endured periods of cold that were too harsh for the horse to have survived locally. The climate argument, based on the spatiotemporal distribution of archeological vestiges, thus seemed to contradict arguments based on the analysis of size profiles, or on the symbolic role of the horse. With such contradictory leads and indications, it wasn't surprising that the enigma of the cradle of domestication had never been solved.

Other arguments based on an analysis of the genetic diversity of contemporary horse breeds also distinguished the Iberian Peninsula from other regions of Europe and seemed to support the thesis of an Iberian domestication. They came from the team of Andrea Manica, based at Oxford University, which had for years already pointed out that this region was a hot spot of genetic diversity within Europe. For this work, Manica studied genetic markers of a particular kind, microsatellites. In a nutshell, they correspond to places on the genome where repeats of very short words, of fewer than six letters each, followed each other several times. A horse could carry on a place of its genome a certain word in a certain number of repeats, whereas a different number could be found in another animal. Rather than through the simple substitution of one letter with another, genetic variability can be shown as a mutation in the number of repeats of the same short word motif. It took only a step to go from that to imagining that genetic diversity measured in breeds from the Iberian Peninsula was greater because it was where domestication had begun, and therefore where mutations had had more time to accumulate than anywhere else. We therefore had to consider this hypothesis too, although scenarios other than that of a precocious domestication could just as well have explained the local abundance of genetic richness observed in this region today—for example, if the region had in its most recent history functioned as a melting pot where breeds coming from a great many regions had converged, or if Iberian breeders had maintained populations on their territory whose size was greater than anywhere else.

The Hypothesis of an Anatolian Cradle

The geographical region that we needed to consider extended from the Iberian Peninsula to the Volga, covering a vast territory of more than 4,000 kilometers. As if that wasn't enough, we had to further extend the field of our investigation so not to omit a final hypothesis: that of an Anatolian origin of domestic horses. In Kirklareli-Kanligeçit, a site located in Anatolian Thrace, at the foot of the Balkans, the horse seems to have made its appearance for the first time right in the middle of the third millennium BCE, at the very moment when, as our data had indicated, there were profound changes in the genetic composition of horse populations. Just as nothing prevented, in the hypothesis of a domestication in central Europe, the horse of Kirklareli-Kanligeçit from being the descendant of ancestors from the north, originating in the Carpathians, we could also tell the history in reverse and imagine that the ancestors of that same horse were native to Anatolia and had managed to cross the Asian shores of the Bosphorus to reach the Carpathians. There is no lack of archeological data indicating that the horse had been cohabitating with humans in Anatolia for tens of thousands of years. It is also in the heart of this same region that many other species of large herbivores, such as the cow, the goat, and the sheep, were domesticated. As long as the vast knowledge necessary for animal husbandry was available on-site, it was not completely absurd to think that it would have been applied to horses as it had been with other animals. At this point, you might be thinking that we could have just as easily hypothesized that the Kirklareli-Kanligeçit horse came from the Romanian shores of the nearby Danube, where horses were known to have existed 2,000 years earlier, or even that it descended from Ukrainian equid populations from the shores of the Black Sea. But without a genetic analysis of the entirety of these populations, all that we could offer were conjectures and speculations about possible origins. The time had come to proceed to the analysis of genomes.

The Answer at the End of the Tunnel

To imagine that what came next was simply a technical formality in which we just had to put samples into instruments that immediately gave us the answer to our questions would be a mistake. It took four years. We had to carry out more than 200 radiocarbon datings and generate several hundred billion sequences to finally solve the mystery of the cradle of domestic horses. Just like the heart of the unfortunate Phar Lap, most of the samples we analyzed simply no longer contained DNA, or didn't contain enough; we had to constantly roll up our sleeves, get back in a plane, and remobilize our collaborators in the field or contact new ones to obtain the material we needed for our analyses. In the end, we had to analyze many thousands of specimens in order to sequence the genomes of 264 of them. Since at the time we were beginning the study we didn't have all the samples that would prove necessary, the huge territories on the

map we had hoped to cover could only be filled in very late in
the project. We advanced small step by small step, week after
week. There was no sudden flash of lightning, no shout of
"Eureka!" But we progressed from a time when we knew nothing to one when we would know everything. It was a long-term
collaborative work, extremely methodical, in which all the members of my team, each at their station, would tirelessly repeat the
same tasks for long months and long years. The final study,
which made the cover of *Nature*, bore witness to what that colossal collaborative effort truly was. It had mobilized more than 160
colleagues in more than thirty countries across the world.

I will never be able to thank them enough, because they had
to stay focused on reading the messages that I sent to them
regularly as new data came in from our sequencers. Most often,
my messages indicated that we needed to look for new specimens from such and such a place, because those we had weren't
sufficient. Sometimes they pointed out that we had to put all
our effort into a very particular set of samples because at the
time they seemed to be the most promising we had ever analyzed. How many times did I believe samples contained the
ancestor of all modern domestic horses? How many times did
I have to correct myself and change priorities that had been
defined a week earlier, in light of new data we'd just obtained?
The people who worked alongside me often heard this refrain:
"You'll see, these samples are the most important we have, by
far"—and then they would prove me wrong almost immediately. I believe this has since remained a joke in the lab. For close
to four years, almost every week, when we obtained sufficient
quantities of new data, I forced myself to repeat the same type
of data analyses so as to motivate the troops and to bring, as
best I could, our ship to port. But in fact, we were most often
sailing in fog.

The Geography of Origins

One of the analyses we really liked enabled us to show each genome as a point on a 2D graph, and to place horses that shared the same genetic history next to each other. Provided we had enough genetic data—that is, that we would have sequenced a not-negligible part of the genome—it was rather easy to say whether a certain specimen was closer to Przewalski's horses and their Botai ancestors, or if it was instead closer to contemporary domestic horses and their well-known lineage, the one we had baptized DOM2. If a new specimen was closer to the former, we knew we had to seek the source of domestication elsewhere, but if it was closer to the latter, we knew we were on the right track. The idea was to continue the exercise and get closer to our target until we hit it.

Beyond this basic principle, luck would have it that as we were analyzing new genomes and adding new points to our graph, a geographical logic seemed to emerge. This was particularly true of the horses that lived before the third millennium BCE; those from France and England appeared genetically closer to each other than they were to those from the steppes or those from Anatolia, for example, which appeared farther apart on the graph. We could therefore read our graph as a map that drew the outlines of large geographical basins gathering horses that were similar genetically. We then understood why the Botai horses were not the ancestors of our current domestic horses: they were quite simply not native to the same geographical basin, and had not blended with horses who originated in other basins. If they had, we would have had a great deal of difficulty distinguishing them.

It required special elements to explain the map of genetic affinities we were in the process of unveiling as the project advanced. This assumed that horses had not undertaken a great migration during their lifetime, but that they had remained essentially confined to the immediate surroundings of their territory of birth. It was also necessary for natural elements to have acted as geographical barriers preventing (or at least hindering) their circulation from one of the great geographical basins to another: large mountain chains, like the Pyrenees or the Caucasus, or bodies of water like the Black Sea or the Caspian Sea, which had also acted as barriers. To create a simple image of the map of genetic affinities among horses from before the third millennium BCE, we could represent populations from the same geographic basin using a unique color and symbol, like a sort of flag whose shade or design would marginally change when going from one group to its neighbors. Change the geographical basin and you will in contrast find a clearly different flag, with its new range of shades and designs.

The Lower Volga-Don Valley: The True Cradle of Our Domestic Horses

As we fine-tuned our analyses, we realized that the situation had remained fixed for most of the third millennium BCE. The same geographical barriers that had prevailed for millennia seemed to have continued to wield their weight to constrain the genetic composition of horse populations. But we were forced to note that another geographical logic had been put in place beginning around 4,200 years ago, when we saw one of the flags displaying a shade and design that were unique to it begin to leave its native region to go to territories it had never occupied, ultimately covering the entire map around the middle of the second millennium BCE. What was this flag's original basin? Did it come from the Iberian Peninsula, from central Europe, or from Anatolia, as some of our hypotheses suggested? Our data didn't validate any of those hypotheses, except one: that of Pavel Kuznetsov, who had one day pointed out to me the place where domestication must have begun in the region straddling the lower valleys of the Don and the Volga rivers and the steppe of the North Caucasus. We had finally discovered the geographical contours of the mysterious region we were looking for: the true original cradle of modern domestic horses.

The set of data we had collected was dense enough for us to be more precise than Kuznetsov had been earlier. We were able to use our data to train artificial intelligence to extract the slightest genetic details, enabling us to distinguish populations in space before the pivotal date of 4,200 years. The goal was to then be able to predict, solely on the basis of the genetic data carried in their genomes, the geographical origin of the horses that had lived since then. The results were striking. Provided they had lived in the last four millennia, each of the domestic horses we had analyzed could only have originated from the same original region: the crescent moon bordering the shores of the Caspian Sea north of the Caucasus. Regardless of the place where they were born, where they had lived, and had been buried—that is, whether they were found on archeological sites from the Atlantic side, the shores of the Mediterranean, or the banks of the Yellow River—all these domestic horses without exception had come from the same ancestors and originated from a single genetic cradle, once circumscribed to a territory of scarcely a few hundred kilometers. Genetics had kept all its promises by providing a precise solution to an enigma that no other previous approach had been able to solve. We hadn't dared to hope for such validation.

Domestication as Reproduction Control

But genetics did even more: it indicated that the geographic expansion of this lineage of horses was accompanied by an unprecedented demographic expansion. It was as if the breeders of the time were very early on able to control and perfect the animal's reproductive activity to produce an ever-greater number of individuals generation after generation. It would be false to believe that horses had taken their time leaving their original cradle and had gradually spread, blending with local populations. Just the opposite was true: they had been reproducing in considerable numbers out of a single cradle and had very quickly overtaken the confines of Eurasia without ever blending with local populations. They spread like wildfire and ultimately replaced the Indigenous populations they met along the way. This means that it was in scarcely one or two centuries that they reached Anatolia, Moldavia, and central Asia, and only a few centuries after that, around the middle of the second millennium BCE, when they arrived in Mongolia, the Carpathian basin, and virtually all of Eurasia.

If we were able to assert that, it is because DNA doesn't just record the geographical origin of horses. It can also provide indications about the effective size of past populations, as we have already seen in chapter 2. Remember that the larger the number of individuals participating in reproduction, the longer the genetic polymorphisms that appear following mutations will stay in a population. The genetic diversity of a population thus gives us a direct measure of its effective size and of its number of male and female progenitors. If we apply this irrefutable principle to polymorphisms of the mitochondrial genome, a genome that is transmitted only from mother to offspring, we can then have an idea of the number of female progenitors in a given period.

By applying it to those of the Y chromosome, which is only transmitted from father to son, we can similarly estimate the number of male progenitors. We were able to affirm that both maternal and paternal lineages, mares and stallions, had experienced a demographic expansion together, one of colossal magnitude; it had on average increased more than tenfold. It seemed that the breeders of the time didn't choose to use only a small portion of horses, but to mobilize all the animal resources they had to produce the greatest possible number of individuals.

In addition, it appeared that the very first breeders had found the means to accelerate the rhythm of horses' reproduction in a very significant way, since our data showed all the signs of a sudden acceleration of evolution beginning in that pivotal period represented by the end of the third millennium BCE. We can be sure of this, because it was from that moment that the genetic distance separating horses from donkeys began to increase more rapidly than it had during the preceding 4 million years, since the moment when these two species separated. If an excessive number of mutations all of a sudden began to distinguish the horse from the donkey, whereas previously there had been an approximately constant number of them generation after generation, it is because new generations had begun to appear more quickly than they had in the past. In other words, the genetic distance that separates horses from donkeys would increase twice as quickly if four years, and not eight, separated two consecutive generations. This is what our data seemed to indicate. They further contained other proof that pointed in the same direction, indicating that the time between generations of horses had been greatly reduced beginning toward the end of the third millennium; we observed that the interchromosomal genetic recombination, the one that occurs

between the chromosomes of the same pair, had also acceler-
ated. This type of recombination can only take place at the moment when sperm and eggs are formed, and we can only see
them emerge in each generation. Finding an excess means that
the generations are increasingly closer together.

How were the breeders of that time able to increase the rhythm
of their horses' reproduction? We don't know for certain, but if
we remember that in nature, a young male must first overthrow
an older and more experienced dominant stallion before hoping
to be able to obtain his own harem of mares, breeders must have
been able to enable their stallions to mate earlier than in nature,
causing breeding that would not have occurred without their intervention. Perhaps they were able to have their mares be covered
(i.e., mated) sooner than they normally would have been, and
create conditions to prevent pregnancies in younger ones from
resulting in miscarriages. Other studies will have to specify the
modus operandi that prevailed. DNA attests with strong confi-
dence to something else among those things that breeders sought
out, which could explain why only this lineage of horses and no
other had ultimately overrun the planet. That thing was related
to their biological characteristics.

The Winning Combo:
Docility and a Strong Back

As soon as it was established that the DOM2 lineage had
quickly and overwhelmingly taken over all others, it seemed
natural to wonder if its genome had something particular about
it that could explain such success, if only in part. We began by
dividing all the individuals we had sequenced into two
groups—the winners on one side and the losers on the other,
so to speak—and looked at each of the positions on the ge-

nomes one by one to determine if one letter appeared more frequently and exclusively in the first than in the second. This simple sequence comparison was enough to identify two places on the genome that differentiated the two groups almost like day and night. The first is localized on chromosome 3, right in the middle of a gene called ZFPM1, whose product intervenes during development in the formation of a type of very particular neurons called serotoninergic of the dorsal raphe, one of the important nuclei of the cerebral cortex. Serotonin is a molecule that enables neurons to communicate with each other—we call this a neurotransmitter. It intervenes notably in the regulation of moods and behaviors to limit risk-taking. If we study lab mice that have been genetically modified to prevent the production of the protein coded by the ZFPM1 gene, the mice show all the signs of anxiety. Placed in cages, mutant male mice will ultimately be more aggressive and bite each other more often than normal. We can therefore reasonably present the ZFPM1 gene as a gene directly influencing behavior.

Going back to our horses, a mutation affecting that gene already existed among wild populations, but it didn't take off before breeders began to accelerate their reproduction to respond to the increasingly widespread demand for horses. Without knowing it, through their reproductive choices breeders favored those that carried that mutation so that it could quickly spread to an overwhelming majority of domestic horses. That form of selection is called "artificial" to remind us that it is engendered by human activity and not by nature. It probably worked, as the mutation affected the horse in a way reminiscent of what was seen with mice in the lab, though in an opposite behavioral way: horses were less aggressive, more docile, and less anxious. In short, their behavior must have adapted more easily from an increased proximity with other horses. It would also be affected by increased proximity with breeders, and with the people who would use them in their travels and perhaps even consumed their milk, if we are to believe the traces of equid milk protein that have been found in the dental plaque of two people who lived in that region in the same period.

The genomes of horses of the modern domestic lineage also differ from the others at the level of chromosome 9. The underlying region involves the promoter of the *GSDMC* gene, the genic platform that enables the cell to launch the production of the protein coded by that gene (called *expression*). In its modern domestic version, the promotor most often contains a DNA segment that isn't found in the others and that very likely influences the level or the conditions of the gene's expression. Studies by a great many medical researchers have shown that those of us who manufacture more *GSDMC* than others run a higher risk of developing a syndrome known as lumbar spinal stenosis, in which the canal through which the nerves leave the spinal fluid narrows. This causes local inflammation of the nerves and chronic pain, and sometimes, when the lumbar region is affected, makes it difficult or impossible to walk. A too high expression of this gene is associated with a fragile back and with painful locomotion in humans. When we see the explosive geographical expansion of DOM2 horses, we can't help but think that a strong back and the ability for movement had to have been useful. It is reasonable to assume that the first breeders of this lineage rapidly favored the reproduction of animals that carried the version of the gene associated with characteristics most conducive to responding to an unprecedented demand for mobility.

In the end, DNA, along with precise radiocarbon dating, had done much more than just help us locate the cradle of domestication of modern horses in the heart of the steppe of the

Lower Volga-Don valley in the northern Caucasus 4,200 years ago. It had begun to reveal the biological reasons for which this horse, rather than another, had become dominant: it was more docile and tolerated close confines better; its back was stronger and was built to endure long journeys without difficulty. Beyond opening the doors to the animal's domestication, the answers DNA delivered also directly challenged the foundations of a theory concerning the history of an entirely different species—our own. Let's now see how.

4

The Horse of the Apocalypse

Indo-European Languages

In France, like more than 3 billion other people on earth and close to half the global population, we speak a language belonging to the family of Indo-European languages. If this language family was so successful, it is because in addition to French it groups together the languages of other nations that have had voracious colonial appetites: English and Spanish. But that is not the only reason for its success. The Indo-European languages had already spread over large territories before the emergence of the great colonial empires that arose during the Age of Discovery and ushered in the modern age, before English became the new *lingua franca* of the world. Latin, for example, is an Indo-European language, but ancient Greek, Hittite, and Sanskrit are no less famous ones. In truth, the list of Indo-European languages would quickly appear interminable if we attempted to refine it with the great scholarly subdivisions that compose it and that are broken down into six language branches that today have fallen into disuse—dead languages—and eight branches of languages that are still alive, some of which are thriving. There are close to 450 Indo-European languages that are spoken throughout the world, more than two-thirds of which form the great branch of Indo-Iranian languages that are principally spread over the Iranian world, the southern Caucasus, Pakistan, India, and Sri Lanka.

If linguists classify languages by families the way collectors inventory their favorite stamps and entomologists their ants, butterflies, and other beetles, it is because they possess common characteristics both in their vocabulary and in their structure, syntax, or pronunciation. For languages to have existed, there of course had to have been people to speak them. The linguistic connections between Indo-European languages therefore most likely reveal the existence of populations that in a very distant past lived in the same region and all spoke the same language; some groups would have later left that cradle of origin and gone to other regions of the world where their language would have been gradually modified over time and formed new branches of languages with common roots. We must see Indo-European languages not as static encyclopedic entities emerging *ex nihilo*, but as having engendered each other throughout history. We can follow that history somewhat like biologists do when they trace back links that led to species living today, or the way genealogists do when they look to familial links, their work resulting in the shape of a tree. It is precisely that tree that linguists have attempted to reconstruct for more than a century and a half now. Their goal is to find the kinship links retracing the history of Indo-European languages and with it that of their speakers, from the original cradle—the *Urheimat*, where the mother of all other languages, proto-Indo-European, was spoken—to discover the very words of this proto-Indo-European language as well as the people who spoke it—the *Urvolk*—and determine where and when they had lived. These are the questions that fuel intense linguistic debates.

But going back along the branches of the tree of languages is not an easy task, for many reasons. We don't know everything about ancient languages, and some that acted as true conveyor belts from one branch to another did not reach us. There were languages that were uniquely oral, before the invention of writing; in other cases no written trace of them has yet been discovered, or, for lack of a Jean-François Champollion and his Rosetta Stone, we haven't yet figured out how to decipher them. In addition, words are not transmitted with the immutable rigor of the laws of heredity that rule biology and the passing of genes from parents to offspring, following strictly familial genealogies; they are borrowed left and right and mix from one family to another, depending on the modes and vagaries of history and sometimes even between groups that have no common roots. For languages, proximity does not necessarily equal kinship, and many models have logically been proposed to understand the history of Indo-European languages. As a result, that history is still the object of great debates today.

Horse-Related Roots

How do horses fit into all that, you might ask? A key model in this linguistic debate places the proto-Indo-European language at more than 7,500 years ago in the heart of a Neolithic Anatolian cradle, while another model situates the *Urheimat* more to the north, on the Pontic steppe of the northern Caucasus, once again in the heart of an *Urvolk*—with horses. We have returned to our favorite animal in this history. Looking at the words closely, in particular those that share the same root among the quasi-totality of Indo-European languages, there was one that was consistently found: the one designating the horse. If we were less demanding and agreed to set aside the Anatolian and

Tocharian languages, only two among the many other branches that make up the tree of Indo-European languages, other words unequivocally associated with the horse, such as "mare," "foal," or "hoof," also reveal common roots. The same is true for words designating the horse's tail, as well as *koumiss*, the alcoholic beverage made of fermented mare's milk that nomadic horse-riding peoples adore, if we consider only the Balto-Slavic and Indo-Iranian branches. Finally, if we concentrate solely on this latter branch, which is clearly the largest because it assembles more than two-thirds of all Indo-European languages, we can find other common roots behind the words "cart," "cart-driver," "bridle," and even "racehorse." This is why it seems reasonable to think that the people who spoke the proto-Indo-European language were familiar with horses and had domesticated them to the point of having used them for riding, harnessing to carts, and transforming their milk into beer.

Horsemen of the Apocalypse from the Steppe

The theory that Indo-European languages originated among steppe horse riders isn't new. It was given new life when the Lithuanian archeologist Marija Gimbutas, who has spent most of her career at the University of California Los Angeles, combined it with a successful narrative, both on the scientific and media levels, at the end of the 1950s. The narrative has been refined many times since then, notably by the famous archeologist David Anthony in his book *The Horse, the Wheel, and Language*, for which he won the 2010 prize from the American Society of Archeology. Gimbutas and Anthony saw in the material and funerary traces of the Yamna culture possible roots of the proto-Indo-Europeans. The people who developed that culture generally buried their dead lying down, curled up, or lying on their side, with ochre placed on the head, in oval or rectangular graves on top of which were imposing tumuli, serving as true burial mounds. It was that same word that described these monumental constructions—"kurgans"—that Gimbutas used to name her theory. The Kurgan people, according to her theory, were a nomadic people who lived on the Pontic steppe from the end of the fourth millennium to the middle of the third millennium BCE. Conquerors by nature and dominant through weapons, and being the first to tame the horse, they had the means for their expansionist ambition. Gimbutas even proposed a map of the temporal and geographical phases of the expansion of the Kurgan culture from its initial home on the steppe into the heart of Europe, where not only did it bring war that resulted in the advent of inegalitarian societies, but also imposed its patriarchal model. By disseminating the apocalypse there, the various Kurgan waves would have deeply disrupted the balance of the old world. In Europe, this change would be visible archeologically in the emergence of new material traces, such as those of the so-called Corded Ware culture. That culture developed in northern Europe in the first half of the third millennium BCE, from Russia to Germany and northeastern France. Among its attributes appeared a new way of burying its dead in tumuli, battle axes of polished stone, and vases embellished with cord-like designs.

Such theories did not wait for Gimbutas. In the 1930s and 1940s, for example, the search for a dominating *Urvolk* had already enthralled fans of Nazi ideology, those avid to find the foundations of the superiority of a people having been able, in the past, to impose its yoke and its language on others (not surprisingly, German is also part of the Germanic branch of Indo-European languages). Even though it only describes a people united by language, *Urvolk* was for them synonymous with ethnicity—in other words, race, obviously that of their prestigious ancestors. They forgot rather quickly that the notion of a people doesn't imply the existence of a united centralized political entity and does not necessarily group together people who are all alike and of a single and same origin as a single and same ethnicity. They also forgot, as others did, that an archeological material unity doesn't equal an ethnicity, either. Who among us doesn't have material objects originating on continents different from our own? But surprisingly, it was in stark contrast to the ultra-right movements, in 1970s California, that Gimbutas's theory made a splash: in the very heart of New Age movements, on a background of feminist militantism adhering more willingly to the patriarchal dominance panel of the theory than to its legendary figures of Horsemen of the Apocalypse.

Follow the Genes to Find the People Who Carried Them, and the Languages They Spoke

Since then, sciences other than linguistics or archeology have tried to tackle the subject. Once again, the science that concerns us here is in the realm of genetics. When you talk about language, you talk about people who speak. You must have people to speak a language, and just as linguists have attempted to climb the genealogical tree of languages, geneticists can try to climb the genealogical tree of genes. To compare the genetic tree to that of languages or to discover in genes the traces of an ancient expansion that in its time had rewritten the genetic and linguistic landscapes of Europe—this is the idea in principle.

This is how the geneticist Luigi Cavalli-Sforza in the 1990s became one of the advocates for the Kurgan theory, because he had been able to discover in the map of European genetic variations striking resemblances with the one that Gimbutas used to describe waves of Kurgan expansion. The coincidence was too striking to be only the fruit of chance, and the genes seemed to vindicate the theories of a great invasion of Europe by a people originating in the steppe, completely reshuffling the genetic maps of that continent. The consequences seemed to have been such that they were still being felt as far as the distribution of genetic variations in current populations: they were massive. The theory seemed to be reinforced some twenty years later when geneticists attempted to use sequencing tools of unprecedented power to retrace the genetic map of populations living in Europe today, but also of those who lived there in the past, notably from the time of the Corded Ware culture and the Yamna culture. It was these same tools that our team used to retrace the same type of maps—this time not from archeological human vestiges, but from those of horses. Once we had these maps, we could replay the film of those migrations and verify the primary expectation of the Kurgan theory: that is, that humans and horses left their native lands on the Pontic steppe together around 5,000 years ago and populated Europe. The map of human migrations appeared first in 2015; that of horses several years later, in 2021. Let's first look at the lessons that seemed predominant if we studied only human history.

Human Migrations Redrew the Genetic Map of Eurasia Around 5,000 Years Ago

These studies revealed that the human populations of Europe had undergone a large reshaping following the arrival of farming and animal-breeding peoples from Anatolia, starting 7,000 to 8,000 years ago. The progression of those peoples was in no way a migratory tsunami, because it seems to have advanced at an average speed of barely 2 kilometers per year. By blending with Indigenous populations, the new arrivals gradually contributed to changing the genetic landscape of European populations so that around the end of the Neolithic period, even if figures vary from one place to another, the humans who populated Europe included about a third of their genetic ancestry that was inherited from local ancestors. The remaining two-thirds took their source into the descendants of the Neolithic Anatolian migrants.

The situation changed again precisely at the moment when archeologists saw the Corded Ware culture emerge. The genomes sequenced from human remains buried in Germany in this context, dating from 4,700 years ago, no longer revealed just the two genetic ascendancies we've just mentioned, but the presence of an additional one—and not simply in trace amounts, as this third ascendancy accounted for close to three-quarters of their genome. It was by looking for other earlier populations in which this genetic component was present in full that we could identify its source, found in none other than the Pontic steppe, in the human remains buried among the Yamna Kurgans. The circle seemed to have been closed, and linguistics seemed finally to have found the Proto-Indo-European people it was looking for. Wolfgang Haak, the author of one of the two studies that reported these results in 2015, decided to title his paper "Massive Migration from the Steppe Was a Source for Indo-European Languages in Europe." For my part, I was a member of the team undertaking the second study whose resulting paper was no less modestly titled "Population Genomics of Bronze Age Eurasia." Our study wasn't limited to Europe but also looked at changes in material culture having taken place at approximately the same time, in Asia, where we

were able to trace the direct genetic links between the populations of the Pontic steppe and others at the foot of the Altai
mountain range. The two studies agreed that the people of the
Yamna culture seemed to have left their native steppe around
5,000 years ago and crossed thousands of kilometers both to the
west (Europe) and the east (Asia). To imagine that the horse
served as their means of locomotion in their momentous journey seemed reasonable.

The studies that followed seemed to support other aspects
of the Kurgan theory, since according to new genetic data the
migration appeared more pronounced on the male side than
the female side. In other words, it seemed that the migrants
were preponderantly males. Moreover, the map that described
the distribution of different Y chromosomes throughout
Europe changed radically at the end of the Neolithic period,
favoring a main type originating from the steppe. It was only a
step from this to accepting Gimbutas's invasion by bellicose
warriors, since the analysis of kinship relationships among
tombs dating from the beginning of the Bronze Age showed
patrilocal links: the people buried were connected by their
father and not by their mother. Women left their community to
join that of their partner, where they ended their days, as their
daughters would do in turn. In this scenario one could see signs
of Gimbutas's patriarchal model. The genetics of the new arrivals from the steppe also indicated that they must have been
taller than other people who lived in Europe at the time,
because among the latter the mutations that in part condition
our height were found more frequently in the right combinations. Perhaps we should see in this one of the reasons—along
with the horse—for their success and their domination. Did
this mean that the linguistic controversies were settled and that
the triumph of the Kurgan model should be acknowledged?

A Weak Link in the Kurgan Theory

I don't believe that, and for a basically simple reason: if the invaders from the steppe used their mounts in combat, and if horses had also provided them means for their geographic expansion, then the genetic map of horse populations should have changed along with that of human groups. The map that Wolfgang Haak had begun to draw in 2015 irrefutably showed that the genetic composition of human groups had already been profoundly modified 4,700 years ago. The upheaval of the genetic map of horses, for its part, occurred only five centuries later, as our work demonstrated in 2021. As final confirmation, the genetic profile of horse remains found in Germany at an archeological site where 4,700-year-old humans were buried with elements of their Corded Ware culture has nothing to do with that of their counterparts from the steppe, but is inscribed in a direct line with that of the local populations of Europe. We had to accept the evidence: the visible tidal wave of humans into Europe was not accompanied by horses. The peoples of the Yamna culture hadn't arrived on horseback, and Gimbutas jumped to a conclusion in designating them as a horse people. It's clear that they were a pastoralist culture, but more likely breeders of cattle and sheep rather than horses. Their vehicle wasn't the war chariot with light wheels to which horses were harnessed but a wagon, with huge, bulky wheels that could only be pulled with the strength of oxen. If the invading marauders with horses lost the instrument of their domination in combat, what must we think of the circumstances of the migration originating from the steppe? Was it ultimately as massive and fleeting as the 2015 paper had described it, and had it really been accompanied by wars that seemed central to the edifice constructed by Gimbutas? Today, we don't think so.

Researchers have used computer simulations to estimate the speed at which these migrations might have occurred to explain the perturbations measured in the genetic maps. These simulations are fairly sophisticated, and attempt to incorporate realistic statistical models of the world in which available resources are modeled by ecological parameters along a grid in which the points of analysis are spaced every 100 kilometers. In these models, populations are allowed to change their demography from generation to generation as well as to migrate from their point on the grid to their neighbors, like migration taking place on foot. They could also jump further away on the grid, at much greater distance, as they would do on horseback; these models can also assume different migration intensities and demographic scales.

Today we don't know what those intensities and scales were, but by allowing computers to simulate what the world would look like for as many combinations of parameters as possible, we can rediscover the figures that are capable of reproducing the genetic maps measured in past human populations. This approach, known by the scientific name of approximate Bayesian computation, enables us to find among necessarily complex models those that are the most likely on a statistical level (I use the term "complex" because these models attempt to explain the mobility of peoples of the past over great temporal and geographical expanses). Even if they are easier to describe than to achieve—simulations can unfold over weeks, even with the help of powerful computers—we must remember one essential thing: they support our interpretations of the genetic data concerning horses.

The most likely models to date suggest practically no ultra-fast migration and return estimations of the speeds of progression of human groups on the order of 4 kilometers per year on average. The migratory wave had been faster than that which had accompanied the farmers of the Neolithic period, but it didn't appear to have occurred extremely rapidly, as previously thought. Furthermore, the best models did not impose competition between migrant groups and local populations for access to natural resources, which seems to rule out war as a massive explicatory driving force. These same models imply demographic declines on both sides—among the new arrivals as well as the Indigenous population, and also including intermixing before a later demographic rebound occurred. This goes to show that one must be skeptical of the great archeological maps when they too hastily sum up the complexity of the whole. Other archeologists, such as Martin Furholt at the University of Oslo, were able to keep a clear head when the genetic maps first appeared, reminding us as soon as the work of 2015 was published that nature and culture—in other words, ethnology and languages or types of ceramics—had no reason to always go hand in hand. To depict the people of the Yamna culture or those of the Corded Ware culture as a great, homogeneous whole, both genetically and culturally, was to jump the gun. It was important to keep in mind regional components before concluding there was a great replacement.

Another irony of history: it was with Wolfgang Haak that I went to Kislovodsk in October 2019, to an old Soviet kolkhoz converted into an archeological museum where alongside half-gutted military trucks, hundreds and hundreds of human and animal bones discovered in the region are stored. That is where some of the vestiges of horses that in our analyses appear to be the closest to current domestic ancestors, those that left the steppe only 4,200 years ago, and not 500 years earlier, are conserved. There are, for example, those from the Aygurski 2 site, associated with the Maykop culture of the steppe. Others, like those from Sosnovka, not far from there, are attributed to another culture, the Poltava. The last belongs to the Repin culture, which some consider to be a cultural entity derived from Yamna. Even if the geographical area where our modern horses were born was circumscribed, our work showed that their inception had not been the prerogative of a single and same people living within a single and unique culture.

The Expansion of Indo-Iranian Languages: A Story of Carts and Horses

The above covers the European side of the population movements that came from the steppe, which brought the seeds of an Indo-European language to the continent; perhaps against the expectations of the Kurgan theory, neither horses, nor horseback riding, nor horsemen were involved. Can the same be said for the Asian side of history, and in particular for the migration of Indo-Iranian languages? Not entirely, and it is the other contribution of our work to show that the same phenomenon—the diffusion of the same large language family—did not always obey the same rules depending on the history and the geography of the place concerned. The genetic maps we have retraced indicate that the expansion of the fully developed domestic horse DOM2 is inseparable from an archeological reality associated with another material culture, that of the Sintashta.

Archeologists discovered the people associated with this culture in tombs that were thirty-eight to forty-one centuries old. The main feature of these burial mounds is that one finds in them the first true chariots or carts, with spoked wheels, vehicles built specifically to be harnessed to horses. Moreover, it wasn't rare for a man to be buried not only with his cart, but also with the two horses still harnessed to it. These are signs of a profound change in history: a moment when the horse, by way of the cart, assumed considerable importance, reconfiguring the maps of human migrations. Human genetic maps indicate that it was at that very moment that new human groups left the Pontic steppe and spread through and beyond central Asia. Here, the temporal and geographic coincidence between humans and horses is almost perfect—the reverse image of what we found in Europe. By domesticating the horse, human groups had found a mode of transportation. And by inventing the spoked wheel they had modified their vehicle, which up to then only oxen could pull, to make it lighter and well-suited for a horse. The combination of these two innovations, one biological, the other technical, would provide the fuel for great human migrations with multiple consequences. Through interbreeding, the newcomers modified the genetic composition of Indigenous human populations along the way. They also brought the foundations for a new language that is at the origin of the more than 500 Indo-Iranian languages still spoken in that part of the world: the proto-Indo-Iranian language.

That great expansion also had consequences for the horse. Breeders of the time controlled its reproduction in order to increase its numbers to an unprecedented extent. Contrary to what happens during human migrations, this new type of horse did not crossbreed much with its cousins—the populations of wild horses—encountered along the way. It always reproduced from the same stock, so that it ultimately replaced almost all other populations, with the exception of Przewalski's horses. There is no doubt that the search for a new means of locomotion, represented by the horse and the cart, constituted a key element in the success of the horse we have called DOM2, the direct ancestor of our modern domestic horses.

But for now, it is not certain that this should be seen as the only reason motivating its domestication. Some have proposed that mare's milk entered the food supply before the incredible expansion of the horse; dental plaque preserved on human teeth found at two sites in the North Caucasus and possibly dating from more than 4,600 years ago seems to contain residue of proteins characteristic of mare's milk. However, more recent work has not been able to replicate these findings, and strongly suggested that horse milking could well have been absent from that region for more than a thousand years after the expansion of the DOM2s. In all likelihood, human groups in the region obtained milk not from mares but from cows, sheep, and goats. If not their milk, could something other than their simple ability to pull carts have been sought in horses? It seems reasonable to respond in the affirmative; insofar as it could be ridden, the horse was able not only to constitute a source of energy—the motor, so to speak—but also the vehicle itself. Among the first DOM2 horses we have found are those from Czech, Moldavian, and Anatolian sites; radiocarbon dating suggests they lived around 4,000 to 4,200 years ago. In those times carts were still unknown in these regions, whereas iconographic sources (such as drawings that appear on a pre-Kassite vessel found in Meso-potamia, Iraq) show a man on horseback. Its age is difficult to assess precisely, but it has been agreed that it dates from around 4,000 years ago. Evidently, the horse would have been used at the beginning of the selection of the DOM2 before spoked wheels and carts—quite simply, for riding.

A Domestication to Confront Climate Change?

Since the answer seems so simple, we have a right to wonder why it took so long before we could domesticate the horse and ride it. Why did this happen only 4,200 years ago, when the techniques for the domestication of animals, other herbivores, had been in use for millennia? Let's beware of drawing facile conclusions; we have seen many times how hasty simplifications could lead us onto false paths. But it so happens that 4,200 years is a date associated with an important climate event that in places might have had disastrous ecological consequences. It was in a paper in the journal *Science* that I read for the first time about the 4.2K event (K for *kilo-year* means 1,000 years, so 4.2K designates an age of 4,200 years). The paper established the temporal link between the fall of the Akkadian Empire of Mesopotamia and a period of repeated drought in that region. Paleoclimate work has since shown that the event was felt well beyond just Mesopotamia; some say it could have led to the lowering of the water level of the Nile and hastened the fall of the Ancient Egyptian Empire, and perhaps even the collapse of the then-flourishing Indus Valley civilization. Specialists are still debating whether what appears to have been a single event does not hide a succession of asynchronous micro-events in different regions, thereby appearing more important and global than it really was. But there is no doubt that a wave of droughts occurred around that date in Mesopotamia, in the Mediterranean region, and over at least part of the Middle East.

What consequences did these droughts have on the North Caucasus steppe? To my knowledge, we don't have the answer to that question. All the same, we can reasonably imagine that if their surroundings no longer offered them sufficient resources, people didn't just lie down and die. They could quite simply have chosen to leave, and it was perhaps at the time of that migration that the horse found its new function. Thanks to the horse it became possible to travel, to connect with territories from which the dry and hostile expanses they were leaving behind seemed to be separated.

Since that time the horse has had many other occasions to influence the historical trajectory of peoples by combining its power with that of other mechanical innovations such as the cart. Today we better understand the importance the combination of riding horses and archers had over the course of history. We owe this understanding to the work of Peter Turchin. The title of one of his recent papers is sensational: "Rise of the War Machines: Charting the Evolution of Military Technologies from the Neolithic to the Industrial Revolution." According to Turchin, that development had the effect of a bomb on the sedentary societies of the period whose economies were agricultural. This was the dawn of the first millennium BCE, around 3,000 years ago. Those societies survived by developing new types of protection, specifically armor that could resist being punctured by the metallic points of arrows; they also invented new weapons such as the crossbow, and developed a new type of organization, supported by an infantry with reinforced numbers. After all, defending oneself with the help of others is often more effective than defending oneself alone in one's corner. The emergence of archers on horseback started a race for weapons that necessitated the placement of ever more complex administrative systems, facilitating the mobilization and control of ever larger armies. The war machine was launched.

Turchin and others saw in this one of the reasons for the emergence of new ideologies exalting a feeling of unity, capable of mobilizing large and cohesive numbers of people acting as a single man in the name of the same cause. Among these ideologies were the great monotheist religions. If Turchin saw this correctly, it would reflect a butterfly effect of gigantic proportions: a small, apparently harmless event, the invention of the metal bit and the bridle, a new means of controlling the horse, would have ended up in the emergence of archers on horseback, followed by that of other weapons and other means of defense, involving ever more massive and sophisticated armies. An endless arms race. And history doesn't stop there, since much later other means of facilitating riding were invented, such as the stirrup, which was first made of leather and then of metal no later than in the fourth century, which would lead to the emergence of new means of combat—the shock between heavy cavalry and their warhorses, which would lead to other deadly spirals.

5

The Horse Before the Horse

A Dive into the Depths of Prehistory

It was an October afternoon. The sky was low, more like it generally appeared in November. In an hour, a guide with a heavy
southwestern French accent would tell us the history of the cave
we were about to visit and the story of its discovery. I was with
my wife and our children; in the back of the museum, we would
just have to go down a few steps and cross the threshold of a
door, leaving the neon-lighted world to find ourselves plunged
thousands of years into the past: in the very heart of prehistory,
in the belly of one of the countless hills in this part of Quercy, a
pech in local Occitan, the Pech Merle. The visit began in a very
simple gallery; although it was enough to remind us that we had
entered a cave more than 10 meters underground, it didn't prepare us at all for what was to come. Very quickly there emerged
before us the claw marks that cave bears had left on the rock
before disappearing forever. There were shapes of almost spec-
tral mammoths humans had drawn there well before the history
of the world began to be written. Then, the footprints of a young
person, fixed in the mud, maybe 25,000 years ago. What was he
doing there—or was it a girl? Were they alone or was someone

66

else around? What were they thinking about, they who knew nothing about us? Only time separated our footprints.

I was lost in these thoughts when finally there appeared what I had come to see, what we had traveled here to find: the panel of dappled horses. The panel was on large block of rock more than a meter and a half high and more than 3 meters wide, featuring two white horses with coats dotted with black spots, their backs turned. To be honest, I didn't imagine they would be so big. I had seen them hundreds of times in photos, but now that I was standing in front of them, scarcely a meter away, it was as if I were seeing them for the first time. I remember being struck when I noticed how the head of the horse depicted on the right seemed to be coming right out of the rock, though it hadn't been sculpted. The artist had chosen its placement judiciously, creating his work on a rock whose natural contours evoked the shape of a horse. Its spotted coat seemed to come straight out of a western, looking almost exactly like the Appaloosas of the Native Americans or like the Danish Knabstrupper, both spotted like Dalmatian dogs. But what exactly were we seeing? Was it a sort of photograph of the horses that had lived there, the way they were at the moment when the artists had painted them? Or were they pure invention, straight out of the artists' imaginations? Were we to consider the prehistoric artist as a naturalist painter or as a precocious surrealist—or should we have sought even other messages in them?

The Place of Animals and the Horse in Prehistoric Art

For a long time, prehistorians have sought in ethnography, or in what these representations and the way in which they are displayed in caves might evoke, elements of a response to these questions. And they are still seeking. Their work has involved observing the artistic practices and the significance of artistic works among the rare hunter-gatherer peoples who still live today in the hope of extracting the reason behind practices that might have been in use in the Stone Age. They have also applied structuralist principles to analyze the place of artistic works within caves and their respective statistical distributions, hoping to reveal hidden regularities that could be clues that would reveal a few snippets of beliefs or visions of the world held by our ancestors. André Leroi-Gourhan, the French archeologist and paleontologist, believed he had solved one of these mysteries when he uncovered in French cave art simple recurrent patterns of association or opposition, such as the predominance of the bison-horse pair believed to represent female and male principles. But others have drawn from these same systematisms exactly opposite interpretations. And others, such as the archeologist Jean Clottes, have steered possible interpretations toward shamanism, noting the overall absence of human figures in favor of animal representations. In his opinion, the artistic work represented the animal auxiliary that enabled shamans to travel and communicate with the world of spirits and obtain their aid and guidance. Therefore, one shouldn't be surprised that the animals represented seemed to float on the wall, without ever standing on the ground or in a landscape: the image referred to hallucinatory visions seen by the shaman during a trance as they crossed mystical realms, and unanchored to the real world and not subject to the laws of gravity. As we can see, even if we have come a long way since the arbitrary representation of art for art's sake, which dominated thinking at the time of the discoveries, there remain many divergent interpretations. We are still at the groping stage in a search for the profound meaning of these fascinating images that have reached us over millennia.

In the past few years genetics has added its two cents to these debates, causing two seemingly opposite worlds to meet. On the one hand there is the world of nature: of what is biologically innate, what is given to us at birth through the genetic lottery. On the other hand there is what is learned: culture or art, the product of experience accumulated throughout one's life and potentially transmitted through nonbiological means. The dappled horses of Pech Merle have assumed a central place in this unexpected collision between these two worlds, because to demonstrate that there could have existed in the time of the Gravettians (around 25,000 years ago) horses with Appaloosa coats might directly change our understanding of cave art. If this were the case, cave art could assume a figurative dimension. While it should not be stripped of magical or esoteric interpretations, those would no longer be the cornerstone that explains the phenomenon. The work of Rebecca Bellone at the University of California Davis has enabled the beginning of an understanding of the biological mechanisms that sometimes cause Appaloosas to be born with a dappled coat.

The Appaloosa Coat

The story involves a single chromosome, chromosome 1—more precisely, it begins with a gene with the scientific name of *TRPM1*. This gene commands the synthesis of the protein also called TRPM1, which plays a role in the conveyance of calcium, notably within melanocytes, the cells that manufacture the pigment of skin. It also intervenes in certain cells of the retina—these cells are called bipolar—that work in concert with rod cells, which are photosensitive and enable us to see even in weak lighting. Bellone's work has revealed that the *TRPM1* gene existed in several versions in today's horses; it happens that some horses are carriers of an insertion of a DNA fragment at a precise point on the gene. Their common attribute: their coat is always at least in part spotted. Because of its significant size (1,378 letters) and its nature (an endogenous retrovirus), the genetic insertion is not without biological consequence. On a cellular scale, it interferes directly with the normal expression of the *TRPM1* gene ("expression" here designates the production of messenger RNA, the necessary intermediary between genetic information and the manufacture of the protein). Rather than facilitating the expression of the gene, the insertion contributes to reducing the stability of the messenger RNA manufactured during the transcription of the gene, which results in a deficiency in the manufacture of the protein in the cells concerned. In other words, this genetic variant leads to an absence of production of the TRPM1 protein in the melanocytes and the bipolar cells. As a result, the *TRPM1* gene appears to be deactivated in horses carrying the insertion of the 1,378 letters; the protein it usually codes is unable to fulfill its function, because it is not produced. A complex process then follows, leading to the agglomeration of melanocytes at precise points on the body (this is called melanocytosis), which results in the formation of the famous spotted coat.

The consequences are not entirely aesthetic, either, since the horses that have inherited the version of the gene containing the insertion both from their father and their mother are also afflicted with what veterinarians call "congenital stationary night blindness," a syndrome that results in a significant decrease in visual acuity under dim light. The deficiency again lies with the missing TRPM1 protein, which cannot exercise its usual role in the transmission of light signals perceived by the rods of the retina via bipolar cells. Because of that, the chain of transmission of light information reaching the retina is lost along the way and the image is never projected to the brain: visual performance is affected, notably under dim light. The Appaloosa and Knabstrupper horses, along with their characteristic spotted coats, also have diminished visual acuity.

Biology connects with prehistoric art insofar as the genetic insertion responsible for the spotted coat has been detected in horses that lived around 15,000 years ago, at the time of the artistic flourishing linked to the Magdalenian culture—the same one that led to the splendors of the caves of Niaux and Rouffignac, also in southwestern France. There is no doubt that horses with dappled coats already existed at that time. Was the same true in the time of the Gravettian culture, close to 10,000 years earlier? The answer is quite likely yes, because another genetic mutation affecting the *TRPM1* gene, which almost always comes together with the insertion described above, was not only present in the Magdalenian horses whose remains have been found in German caves, but also in horses from southwestern France of the same period. The discoveries did not represent isolated cases. Among the sites concerned is that of Igue du Gral, which is a natural trap in which a great many animals perished, seeing only too late the ground disappearing under their hooves. It is located in the territory of the village of Cabrerets in the Lot, a stone's throw from the cave of Pech Merle. The age of the horse carrying the genetic mutation indicates that dappled horses existed in the region 16,000 to 18,000 years ago. There is no reason to imagine that they didn't exist a few millennia earlier, at the very time when the painting was done, especially since we have also found the same mutation in much older horses, as far as Siberia.

Nevertheless, it is clear that nothing enables one to draw the general conclusion that cave art was uniquely figurative; examples of cave representations as supernatural as they are abstract abound. But it appears just as futile to systematically reject out of hand any attempt at a figurative interpretation of certain aspects of cave art. The coincidence of the presence of dappled horses in nature and in the artistic register contradicts any argument reducing horses to a necessarily hallucinatory vision of the world. Furthermore, it is remarkable that among the six horses of Igue du Gral we have analyzed, only one carried the mutation in question and was dappled. That the artists' choice would focus on a rather rare characteristic is perhaps not trivial, in particular if that preference had been confirmed elsewhere, which genetics also enables us to investigate. This might then reveal other patterns that had up to now escaped structuralist approaches. Those classifications could offer new paths to an understanding of what the horse truly meant to prehistoric people and what they sought to express through it, and beyond that, through their art in general. There could be an entire program of research in which the horse would hold a central place, given how frequently the animal is represented in Western European cave art—a program that could be expanded to understand prehistoric practices far beyond it.

The Art of the Hunt

The horse was also among the species that were once the most hunted. Its remains, as well as those of reindeer, are particularly frequent in Magdalenian sites in the Parisian Basin, such as Étiolles, a site that Olivier Bignon, a researcher at the French National Centre for Scientific Research (Centre national de la recherche scientifique, or CNRS) and at the Université Paris-Nanterre, has studied from every angle to understand the circumstances that led to the formation of a pile of horse skeletons in a precise place in the site. The bones didn't show markings characteristic of the activity of carnivores, which would immediately suggest that the accumulation was connected to human activities. The size and the signs of wear on the teeth revealed that the remains belonged to at least three individuals: the first was around nine years old, the second around five years younger, and the last one was a one-year-old juvenile. This seemed to indicate that the Magdalenians had not set off to hunt blindly, but had adopted a well-oiled tactic of collectively organizing to target and kill a family group. In this case, it was most likely a mother and two of her offspring. In a natural state, horses are remarkably social animals who live in herds of several mares with their young and who generally remain under the protection of one stallion. Only young males—the famous bachelors—venture out, leaving the group around the age of four or five to lead a single life for many years until they are able to become the head of a group by dethroning the aging dominant stallion. (Let's note in passing that to speak of a "dominant stallion" as is customary is a distortion, and not entirely fair to the females; it is not unusual that mares are the ones who impose their authority on the rest of the group.) The theory that the familial group had been victim of a hunt made some sense, but it deserved to be tested more formally.

To do that, genetics was again a precious help; it enabled us not only to identify the sex of the individuals—we just had to determine whether they were carrying X or Y chromosomes—but also to establish kinship relationships that would connect them. This will surprise no one; after all, aren't the most well-known genetic tests the ones that aim to establish paternity, or to verify that an individual is related to a very precise family? Genetics didn't disprove Bignon's thesis. On the contrary, it reinforced it: the remains at Étiolles were ultimately determined to have belonged to four individuals, three of the four were females, and were all closely related. It wasn't a matter of isolated bachelors, but a familial group. Far from being anecdotal, the pile of bones in Étiolles offered an important lesson. It provided proof that the taking of familial groups was among the hunting tactics of the Magdalenians. The principle having been established, it was then a matter of repeating the experiment elsewhere and in other archeological contexts to establish if it was a preferential strategy or if the Magdalenians, the Gravettians, or groups from other societies had other methods to use depending on variables such as the seasons, locations, or the weapons they had. The program is now underway, and we are hopeful that very precise elements of a response to our questions can be reported in the years to come.

The Painted Horses of the Chauvet Cave Are Not Przewalski's Horses

The genomic data we have collected have already shattered the most widespread beliefs on the nature of the wild horses that existed in Europe in prehistory. This involves Przewalski's horses, once settled on the Mongolian steppe, as well as the "panel of horses" from the Chauvet cave. This panel is a pure marvel of Paleolithic art, showing four horses lined up in profile. Their necks are massive, their heads high and tapered, and are bulging with powerful masticating muscles; their manes stand straight up, short and black. Feature by feature, they resemble Przewalski's horses, for a long time considered to be the last wild horses living on earth. That is misleading. I was a believer when I photographed these horses in the Seer reserve in western Mongolia, where the TAKH association launched a program to reintroduce Przewalski's horses in the mid-2000s. I went there to take samples of horse manure to compare the bacteria in it with those that are found in animals that live in zoos their entire lives. (We did find rather fundamental differences that should be taken into account to assist in the success of animal conservation programs. But that's another story.)

The horses of the Chauvet cave support the illusion that Przewalski's horses had lived in the Ardèche Valley 33,000 years ago, but DNA does not deceive: those horses don't have much in common. They represent lineages—populations, if you prefer—that had been separated for at least 50,000 years. While we haven't sequenced the genome of the horses whose remains are still preserved in the sediments of the Chauvet cave, we have sequenced those of Étiolles and Igue du Gral, as well as those of Pincevent and Closeau in the Marne basin, those of Kents Cavern, the oldest Paleolithic horses in England, and those of the Goyet cave in Belgium—all contemporaries of those of Chauvet. Their genomes are all based on a common genetic foundation, revealing an evolutive proximity and common origins. That foundation presents no affinities with Przewalski's horses, nor even with their Botai ancestors. Despite appearances, we must accept the evidence: the horses of Chauvet and Przewalski's horses are two very different types of horses.

The Unexpected Diversity of Prehistoric Horses

Prehistory isn't limited to just the horses of Chauvet and of Przewalski, and our work also reveals the incredible genetic diversity of horse populations before domestication. Imagine: in 2015, by sequencing the genomes of three horses that lived in Siberia respectively around 5,000, 16,000, and 43,000 years ago, we expected to find ancestors of horses living on earth today. If we relied on what we knew back then, it couldn't have been otherwise. The idea was even to build on the differences observed between the versions of genomes carried by the oldest of these three horses and those of the most recent, and those of contemporary horses to identify the biological changes that accompanied the recent evolution of horses, including domestication. For us, it was a useful way to replay the film *a posteriori*—to catch evolution red-handed, in a sense. At least, that's what we thought. Imagine our surprise when we discovered that none of those Siberian horses was a direct ancestor of horses living today! At best, we could see them as distant cousins that have disappeared, members of a lineage that up to then had remained unknown. What we learned from the sequencing of those three genomes is that 5,000 years ago there existed a horse of a third type. It was given the name of the region where it had been discovered: *Equus lenensis*, the horse of the Lena, referring to the huge river that flows from Lake Baikal on the edges of southern Siberia to the Laptev Sea in the north, passing through Yakutsk, the capital of the region. Imagining a human equivalent, it was almost as if we had discovered that a Neanderthal lineage had lived alongside *sapiens* almost at the time of the first pyramids in Egypt. You can imagine how astonished we were.

We didn't yet know that as we continued to decipher the genomes of ancient horses we would go from surprise to surprise—because a fourth type was soon to follow the third. This one didn't come from the most isolated confines of our world, but from lands near France: Spain and Portugal. It wasn't among the oldest, since one of its most recent representatives went back to the very beginning of the second millennium BCE; it was dated at a bit less than 4,000 years. The number of letters distinguishing the genome of the Lena horse and those of modern domestic horses indicated that these two lineages had been separated for around 130,000 years. The letters separating the latter from the Iberian lineage revealed an even more distant divergence: at least a half-million years. We discovered that in the heart of Europe there had existed wild horses of at least two types: those that had evolved on the southern slopes of the Pyrenees, or Iberian horses, which we named IBE horses for lack of greater inspiration; and those located on the other side of the Pyrenees and which we have already mentioned on the subject of Pech Merle, Chauvet, and Étiolles. The latter were similar to other prehistoric horses in France, Belgium, and England, but quite distinct from Przewalski's horses. From the two types still living today—domestic breeds and Przewalski's horses—we went to three lineages with the Lena horse, to four with the IBE horse from the southern Pyrenees, and then five types, with the one from the northern Pyrenees.

The list only got longer each time we enlarged the perimeter of our study, and we quickly stopped counting. Going back to Eurasia, these type variations caused a very clear organizational logic to appear: they encompassed the great geographical basins of Eurasia. It was as if every region delimited by large natural boundaries—chains of mountains (the Pyrenees and the Caucasus), maritime expanses (the Black Sea and the Caspian Sea)—had ultimately circumscribed within its boundaries populations of horses, each giving birth to its own lineage. Inside each zone, two horses found on neighboring archeological sites on the geographical plane appeared closer genetically than two horses found at sites separated by a greater distance. This phenomenon is known to geneticists by the term "isolation by distance." It occurs when natural contacts between populations (i.e., interbreeding) occurs in close confines rather than on large geographical scales. In the opposite case, we would see populations that resembled each other over much greater distances. Here, then, are the great principles that organized the world of horses before domestication: geography and natural barriers.

Against all expectations, and given the fact that it is precisely its ability to take us over very large expanses that earned it its success among humans, the wild horse was rather a homebody and preferred to remain near its natural cradle. Over many generations, different populations evolved in isolation and on the scale of Eurasia, ultimately presenting a remarkable diversity of types. It is for that reason that we were able to discover the cradle of the domestication of horses. If there had existed only a single type wherever it might have been, the genetic point of departure would have looked the same at every point on the planet. Sequencing millions of genomes wouldn't have improved our ability to distinguish them. We owed our success to the geography of Eurasia and to the natural behavior of the horse, which turned out to prefer keeping contacts among neighbors, rather than making long-distance friends.

The Extinction of Prehistoric Horses

At the time these domestications were occurring there was a horse from Iberia, a horse from the northern Pyrenees, a horse from the Carpathians, a horse from Anatolia, a horse from the Pontic steppe, a horse from the steppe of central Asia, a Lena horse, and many others as well. What has happened to them since then?

Though the Przewalski's horse descends from the domesticated horse from the central Asian steppe and managed to return to a wild state, most of the other types died out, except for the horses of the Pontic steppe, which were domesticated and spread across the entire surface of the planet. There are no Lena horses today; the last specimen we were able to identify lived around the end of the fourth millennium BCE. Those that are portrayed on the walls of the Chauvet or Pech Merle caves no longer exist either; we have also lost trace of the Iberian horse soon after the second millennium BCE. This generalized decline follows the natural demographic tendency of diminishing horse populations that paleontologists have been documenting for decades now, because horse fossils became increasingly rare soon after the last glacial period. It is even because of this decline that some people have suggested that if horses hadn't been domesticated they would have ultimately died out and joined the ranks of species that have been lost forever. This would be what has happened in the Americas, if we refer to what most of our textbooks say.

However, some myths remain quite rooted, and there are many places in the world where local horse breeds are believed to represent the last survivors of wild populations otherwise lost—a source of pride for defenders of the local land. The Polish Konik is perhaps the best known of these, since it is commonly described as the direct descendant of the illustrious tarpan horse that once populated the Pontic steppe and that of Eastern Europe. The reality is less glorious. The Koniks living today are modern agronomic creatures derived from a restricted cluster of a few individuals and a rigorous and systematic process of selection aiming to approach as closely as possible the aesthetic characteristics of the late tarpan. Legends have a tough life.

We had the opportunity to sequence the genome of a tarpan specimen that had been captured in 1868 on the Kalmykia steppe not far from Kherson, in Ukraine, and which since then had joined the collections of the Saint-Petersburg Museum. Its genome leaves no doubt: it is a crossbreed composed of around two-thirds of an ancestry of a wild horse that populated

Germany and its surroundings in the third millennium BCE, and one-third from a modern domestic ancestry. The genome of Polish Koniks comes entirely from a modern domestic ancestry; it lacks a good two-thirds of an ancestry found in the tarpan. Consequently, Polish Koniks are not the descendants of a famous tarpan that somehow miraculously survived, alone and isolated, in the haven of peace and protection provided by the Białowieża Forest. Furthermore, the consanguinity of Koniks is far from being anecdotal; because they were reconstituted through crossings between relatives starting with a small nucleus of animals all resembling each other, consanguinity was multiplied by a factor of 100 between the 1950s and the 2000s. If our work demystified the true status of Konik horses, it also taught us that a part of the genetic composition of wild horses that once lived in Europe did not disappear immediately following the explosive expansion of domestic horses around 4,200 years ago. It had survived until the beginning of the twentieth century not in the form of a pure strain but in the form of crossbreeds, the fruit of the mixing of immemorial ancestors and the domestic newcomer. Their complete, definitive extinction didn't take place until quite recently, in the course of modern time.

A Return to the Wild

This nuance has its importance well beyond those of us attempting to decipher the world as it once was. It has practical echoes in today's world and might even have resonance in the that of tomorrow, given that conservation biologists have adopted the mission to remake Europe into a wilder region than it has become in the course of history. This is the motto of the Rewilding Europe coalition, which plans to institute protected zones where nature can reclaim its rights and where communities of species can reestablish natural balances without the intervention of human activity. The hope is that biodiversity and the multitude of benefits that natural ecosystems offer us—such as clean water and insects carrying pollen from flower to flower—will be restored and preserved.

In this vision, certain species are known to exercise an essential role due to the central place they occupy in ecosystems, by reason of the diversity of their activities and the interactions they maintain. They contribute to the resilience of a zone much more than their numbers would lead us to believe. True cornerstones of the system, they attract the very particular attention of rewilding specialists. The neo-aurochs and the Polish Konik horse, ersatz species recently bred to resemble the wild ancestors of European cows and horses, were believed to play an important role in the rewilding experiment at the Oostvaardersplassen nature reserve in the Netherlands. This is a humid zone of 56 square kilometers located in a suburb near Amsterdam that was reclaimed from the water in 1968. While grazing on the vegetation, neo-aurochs and Konik horses were expected to thrive when the place was opened and contribute directly to maintaining a diversity of species other than those associated only with forest ecosystems. Thirty years after the reserve's classification as a protected zone, however, that initiative resulted in a failure that was as bitter as it was tragic. Because populations of large herbivores could multiply there, they experienced terrible famines when natural plant production, necessarily limited in a reserve of constrained space, was no longer sufficient to feed them, and when the winters were the least bit harsh. Close to 500 horses died of hunger during the winter of 2017–2018, and almost six times as many deer died in that same season. Since then the zone has been converted into a reserve where tourists can experience an imaginary pre-civilization landscape, but where the growth of the population of herbivores is regularly controlled through slaughtering and the meat is sold by the agro-food industry.

Whatever one's opinion, the Oostvaardersplassen experiment wasn't just a failure; it enabled us to learn what mistakes need to be avoided if we want such projects to succeed. In particular, it revealed the rule of the "four Cs" as an indispensable key for the success of rewilding projects: natural Corridors must always enable contact to be maintained with Core zones; Carnivores must then be permitted to naturally regulate the number of herbivores, if we don't want to end up having to resort to slaughtering policies devoid of Compassion, and thus ensure the quality of life of the animals living in these reserves. Reality is complex; as reasonable as that rule might appear, it often remains difficult to predict how situations will evolve over the long term.

Look, for example, at the brumbies, those horses that live in semi-freedom in Australia. One cannot say that there isn't enough space on an island the size of a continent—nor enough predators or danger, either. Furthermore, the zones in the center and on the periphery are in contact: three of the four Cs are in a certain sense covered in advance. However, faced with the devastating fires of the past few years, many local ecosystems have been reduced to ashes and are directly threatened. Once the flames were extinguished, the brumbies returned and their trampling and grazing began to threaten the survival of young shoots and the reemergence of vegetation, and ultimately the survival of unique environments. When we know that a horse can spend 12 to 17 hours a day grazing, and that the natural park of Kosciuszko, the largest in the country, saw its population of horses go from 6,000 to more than 20,000 between 2015 and

2019, one can easily imagine the consequences. Australia didn't even have any horses before the species was introduced when the island was discovered by European explorers in the seventeenth and eighteenth centuries. The flora there is not very well equipped to withstand the perturbations horses bring, whereas it is naturally resilient to fires, which are common there.

For the time being, one of the political responses to this situation has been to undertake the massive slaughtering of the animal—now considered invasive—so it will stop imperiling a biodiversity without equivalent on the planet. The decision was confirmed in May 2020 by the Federal Court of Australia, which pointed out that the raison d'être of natural parks is to preserve the native wild fauna and flora and not that such as horses, have been introduced. It seems that the fourth C of the rule is not always the easiest to put into place, in particular when species compete with each other for their survival. And Australia is no exception. In Canada, in Alberta and British Columbia, horses in freedom are causing just as much damage due to their increasing numbers, grazing, trampling, and the resulting erosion of the soil. Here, too, the answer has often been to slaughter them or capture them for resale elsewhere in the country, and some even toy with the idea of chemical contraception campaigns targeting the wild fauna, which could lead to even worse consequences in the event of errors in dosage or targeting.

It would perhaps be logical to identify zones that are naturally compatible with megafauna, where we could then hope to reintroduce the horse at lesser risk. One of the approaches that seems promising consists of guessing the contours of the species' natural ecological niches in order to find zones for them that are compatible with the contemporary world. To do this, one would need to go back in time and compare the maps of fossil distribution to the horse or other candidate species to be reintroduced with those of multiple paleoclimate variables, such as the average annual temperature where the fossils were found or the average annual precipitation, among many others. Current climate models enable us to reconstruct these parameters with astonishing precision. In this way, we would be able to predict which regions in the world would be able to offer living conditions naturally favorable to the species we're interested in.

As seductive as it may be, this approach in the horse's case nevertheless shows obvious weaknesses. As we've already said, for periods before domestication, the distribution of fossils demonstrates a diversity of horses that have essentially all disappeared today. For those that have survived, the fossils suggest movements of humans on horseback more than the ecological contours that naturally suit horses. All the same, despite these limitations, research teams propose reintroducing the horse into certain regions of Scandinavia, such as Denmark, and their choice seems again to veer toward the Konik horse, whose origin, as we've said, is not very wild.

For lack of anything better—that is, for lack of a true wild strain—it would perhaps be best to focus on domestic semi-feral varieties: those that are left free but that lived alongside humans and have adapted to local territories for centuries, even millennia. Such is the case of Dartmoor ponies, originally from the southwest moorlands of England. The idea would be all the more opportune in that there remain scarcely more than a thousand of them at the time of this writing, whereas there were ten times more in the 1960s. It is one of the possible consequences of Brexit and of the ambitious new agricultural policy the Bureau of Land Management claims to be putting into place. In theory, it involves supporting farmers and breeders not based on the amount of land they cultivate or occupy but instead based on their production and the services they render, such as reducing air and water pollution or preserving biodiversity. Moorlands like those of Dartmoor are overpopulated with particularly tough species of plants that neither cows nor sheep will graze. Exclusively maintaining cow and sheep breeding has inexorably resulted in a surge of such plants, leading to an impoverishment of biodiversity, and has increased the vulnerability of the ecosystem as a whole, particularly in the face of fires. Removing sheep and cows wouldn't be a good option for the plants or the breeders either, because other plants would then dominate the land. The solution seems to involve a triad in which cows, sheep, and horses cohabitate in balance and maintain not only the moorlands of Dartmoor and the ecosystemic services they contain, but also a traditional knowhow and way of life.

Recent experiments offer reasons for hope: salt blocks have been placed at strategic points in the area in the hope of attracting horses, which consume close to 50 grams of salt per day. In scarcely three years, the regular grazing of these places by horses, who don't need to be begged to eat the moor grass, was enough to enable the heather, a species typical of the moors, to come back. This is a process we need to follow closely and perhaps adopt elsewhere, if in time the results are confirmed. For now, it shows us how the horse and the relationship we would hope to maintain with it might continue to wield true political weight in these times of crisis in climate and biodiversity.

6

The Other Horse

The 1976 Great American Horse Race

On May 31, 1976, a breath of spring was blowing over Frankfort, and beyond that little city in New York, over all of the United States. The Watergate scandal had brought down Richard Nixon two years earlier, and the carnage of the Vietnam War had come to an end in April the year before. The mood was carefree, and Americans were getting ready to celebrate on July 4 the bicentennial of the signing of Thomas Jefferson's Declaration of Independence. The Great American Horse Race would give an audience of horse racing fans the dreamed-of opportunity to focus on something else. For the more nostalgic among them, the event would offer the ideal pretext to rediscover their love for the Great West and admire the immensity of breathtaking landscapes. For the participants, it would involve racing over nearly 6,000 kilometers on their horse, in part following the former route of the Pony Express to California. They would start on the East Coast, in Frankfort. At the finish line in Sacramento, on the other side of the continent not far from San Francisco, a hefty prize of $25,000 (which today would be worth four times that) would await the winner. Those in the know agreed that the race would not be kind to the more than 200 brave riders who were to line up on the starting line, each with two horses: the primary one for the race and an emergency horse, in the event things went badly. The horses were mostly Arabians, chosen for their legendary endurance, and Icelandic ponies, whose stolidity in the face of the icy expanses of the Great North had earned them the deserved reputation of being invincible.

The race's structure was similar to that of the Tour de France, although shorter stages of around 60 kilometers were ridden day after day. There were additional differences: the horses were regularly checked by veterinarians, sometimes every 16 kilometers, and the doctors could impose a forced rest stop on teams in the event of clearly dangerous fatigue. This didn't prevent many competitors who started off too quickly from breaking down, but after more than 315 hours spent in the saddle—the equivalent of 13 days and 13 nights nonstop—Virl Norton crossed the finish line. He wasn't actually first, but after tallying the penalties accumulated by the competition who were inclined to take shortcuts or were unconcerned with their horse's condition, he was crowned the grand prize winner. To everyone's surprise, Virl Norton didn't ride a horse. Instead he rode a mule. Its name, Lord Fauntleroy, would go down in legend, as would Virl's emergency animal, a female mule named Lady Eloise who was wounded before the end of the race. Team Mule had beaten the most resilient and toughest horses in the world. It was almost unbelievable.

Mules: True Forces of Nature

As for Virl Norton, he wasn't at all surprised. A worker and farmer's son from Wyoming, he knew the qualities of this animal very well: it was half-donkey on its father's side and half-horse on its mother's. He also knew that others before him had taken advantage of the strengths of this unnatural hybrid. As far back as antiquity, the Greeks and Romans had noticed what biologists today refer to as hybrid vigor. The hybrid brings together the qualities of each of its two parents while discarding their weaknesses; using contemporary language, we might think of it as being *augmented*. The mule is a tough animal without peer in attacking a task without ever giving up—which Virl's crazy race had demonstrated. But there is more. The hybrid is more sure-footed than the horse, especially in difficult terrain such as mountains, because its hoof is much tougher. This is the reason why caravans of this animal joined combat zones in Afghanistan to supplement the logistics of transport and provisioning in particularly difficult terrain. It is slower than the horse, but it requires less care. It is also stronger and withstands thirst, hunger, illness, and insect bites better. It is also more patient. These are qualities inherited from the donkey, and mules surpass it in height without inheriting the donkey's well-known downside: stubbornness. For those who have gotten to know it, the mule defies the common image of it as a stubborn, uncooperative, and impossible animal that persists in our collective imagination. Who among us has never been called "mule-headed"? It is true that the animal's ears, shorter than those of its donkey father, are still much longer than those of its mare mother. And let's not forget that the traditional dunce hat in France is adorned with two donkey ears.

The expression "to be loaded down like a mule" is much truer of this animal than any claims to stubbornness, and its unequaled strength has contributed to building empires and has enriched many entrepreneurs, including those in the early years of the United States, which owes it a lot. George Washington was one of the cognoscenti: he became the first American mule breeder. He had been able to obtain excellent Andalusian donkey studs through a direct request of King Charles of Spain, at a time when exporting donkeys was prohibited. This was a royal gift that allowed him to have a leg up, so to speak: barely fifteen years later, there were close to sixty male and female mules that could be seen grazing in the fields of the presidential estate of Mount Vernon. Once the example had been set by Washington the idea would quickly take off, and at the beginning of the nineteenth century it was estimated that almost 1 million of these hybrids were trampling the American land.

Initially the mule was most successful in the southern states, where it was known as a do-it-all animal for plowing, harvesting, and also for transportation. The animal also aroused a certain interest in the West, playing a role in the success of the Old Spanish Trail, a route that was considered one of the most difficult, winding, and longest ever traveled on the continent. The mule had its moment of glory in the 1830s through the 1850s. It enabled entire cargos of wool blankets woven by the Indians to be transported from New Mexico to the California coast. To do that it was necessary to cross deserts that were considered the most hostile on the planet where the animal did not just serve as a beast of burden, but also as currency. Chroniclers report that the Native Americans were also aware of its value, despite their enduring passion for the horse; they could trade two blankets for one horse, but they were prepared to offer much more when it was a matter of acquiring a mule. It didn't take them long to appreciate just how strong, reliable, and resilient a worker that animal was. With the development of the railroad and other more direct routes that soon enabled the conveyance of greater volumes of merchandise over the same distances with less effort, the mule's popularity waned. But its qualities were still valued, and it would quickly be put to other, no less important, uses: in mines, gold mines especially, but also with the US Army, where it constituted an indispensable logistical link for supplying the troops, moving war apparatuses, and repositioning the heaviest equipment.

In Europe, including France, mules have performed many services, because mule drivers are able to pack more than a hundred kilos onto their animals' backs. Until the nineteenth century, entire caravans with their cargoes of salt, fish, wine, and other products continued to travel through the French countryside, starting at the coasts of Languedoc and joining the plateaus of the Margeride mountains north of Aubrac, following ancient routes and passages older those used during the rise of the Roman Empire. When Hannibal traveled through the Pyrenees and the Alps accompanied by his famous elephants in 218 BCE, he also had mules at his side.

Hearty—but Sterile—Hybrids

If there is no end to the praises heaped on this half-donkey/half-horse, it does have one defect that is well known to breeders: it is almost always sterile. It is usually impossible to mate a female mule with one of its kind in the hope of seeing a baby mule born eleven months later; in the overwhelming majority of cases the fetus is aborted before then. In 2013 the San Diego Zoo reported success in bringing a baby mule into the world, but it was by crossing the prodigy's mother with a donkey and not a male mule. The crossing thus didn't result in the birth of a true mule, but a half-mule. It would be necessary to repeat the operation dozens and dozens of times before getting a fertile female mule and enjoying a new success. Any breeder who might attempt to specialize in the production of hybrids from hybrids would immediately go bankrupt. There are not breeds of mules in the sense that there are breeds of horses and breeds of donkeys—that is, animals prized and selected for their biological characteristics that can be reproduced *ad libitum* from parents of the same type. This is an important problem both for breeders and for owners, given how inexhaustible the services are that the animal can provide.

So how in the world did breeders in the past manage to obtain mules in sufficient quantities? One thing is certain: they must have maintained ample herds of each type of the two parents. In those conditions, breeding mules must have cost them a great deal. In addition to mares, they also had to take care of donkeys. Breeding just one or the other of the two parents would seem much more straightforward, unless their mule production was incredibly lucrative or they had managed to overcome the problem by using only a limited number of donkey genitors. That choice would have clearly delighted their bankers, if I may say so, because their investment to maintain the paternal lineage indispensable to the crossing would then have remained limited, whereas their returns would have been greater. They would still have had to choose the genitors carefully, especially since the crossing of a mare and a donkey is not the easiest thing to accomplish and requires a certain know-how—even an art, some would say.

In that case, what were the criteria breeders of the past relied on to guide their choice? Did they select animals that worked harder than others? Or did they prefer heftier animals, in the hope that their offspring would be more vigorous? Perhaps they maintained a herd of horses, castrating most of their stallions so they wouldn't risk impregnating their mares, and bought stud services from their donkey-breeder colleagues in the spring to externalize the costs of maintaining the studs (thereby pleasing their creditors). I was thinking about all of this, and a lot more, when I started to work with Sébastien Lepetz and Benoît Clavel, two of my colleagues at the CNRS, researchers at the French National Museum of Natural History.

Ancient Mule Husbandry

I had already been working with my two colleagues at the museum for a few years. But this time, when we decided to sift through the genetic material of more than 1,200 archeological equid remains, we were clearly shifting gears. We were hoping to understand how breeders of the past had managed and even built up their equid resources of horses, donkeys, and their hybrids in France, and over more than two millennia. It would have been quite an undertaking. The task that confronted us was colossal as well. The COVID-19 pandemic didn't make it easier; Lepetz and Clavel nonetheless managed to take advantage of periods between two lockdowns to visit archeological collections and send us precious bone and teeth samples. As for us, at the lab we organized a team of three technicians, soon joined by a fourth, who used those same periods of freedom to extract DNA from the samples and sequence it.

To determine whether archeological remains come from a horse, donkey, or mule isn't an easy task. Although the premolars and molars of donkeys and horses can appear in shapes that specialists can learn to distinguish, and which seen from above and from a distance are not dissimilar to butterfly wings, those of hybrids are not really easy to spot unless one uses a sophisticated set of morphological analyses. The operation always includes a risk of error in a not-negligible number of cases, one out of ten or even more, depending on the samples available for analysis. We obtained more satisfactory results—an error rate of only 7 percent—by using remains other than teeth, such as the bony labyrinth of the inner ear, whose shape and size we have studied with the help of very advanced medical imaging techniques. During the pandemic, of course, we obviously weren't going to be so indecent as to ask overburdened hospitals to use their scanners to analyze archeological remains. Furthermore, many of the samples we had were not teeth or bones of the inner ear, so we didn't have much choice. We had to move on to DNA to know which of the two species or their hybrid we were dealing with.

Fortunately, in 2017 we had developed a suitable method that was almost infallible, and its principle was simple: by sequencing the genome of each of the living species of equids, we had noted there were many genetic differences. There was roughly around 1 percent from one species to the other, which adds up to several tens of millions of letters where both versions of the genetic information differ. The DNA sequences that we typically find in the fossils contain around fifty letters. For each pair of equid sequences characterized, we had a chance of finding a letter that, depending on whether it carried a given mutation or not, would tell us the name of the species we were analyzing. By not limiting the experiment to two sequences but by repeating it many times, the analysis would quickly deliver clues concerning the species we were dealing with, and if we were analyzing enough sequences, we would certainly know that. A few thousand sequences were enough to reduce to zero the risk of diagnostic error, including in hybrids, because in their genes they carry the mutations of their two parents in equal proportions. As a bonus, the method even informed us of the sex of the animal being analyzed while counting the sequences of the X chromosome issued (and not the characteristic mutations of the species). In proportion, they should appear as frequently as those of the non–sex chromosomes in females from whom the

X chromosome comes, like others, in pairs. And they must appear twice as less frequently in males, because they carry only one X chromosome (the other being replaced by the Y chromosome). Provided that traces of DNA from an animal dead for centuries, even millennia, are still present in the archeological remains, we were then armed with a powerful tool capable of revealing where and when the breeders of the past had raised horses, donkeys, and mules, and in what proportions. It also told us whether a site had a higher concentration of females and had thus perhaps served more for foaling. Since our method had been proven, including for hybrids, we easily found a name for it: we called it "Zonkey," from the name of those animal curiosities, also sterile, that zoos are able to produce by letting zebras into the same pens as female donkeys.

The Golden Age of Mules in France

Perhaps the most striking result of this work was that it revealed that breeders did not always produce the same species. On the contrary, they adapted the nature and composition of their breeding practices depending on the function they required. Horses remained the most popular throughout the twenty-five centuries we studied; in France they represented around two-thirds of equid production. Donkeys didn't appear in significant numbers until the end of antiquity. The golden age of mules occurred between year zero CE and the end of the third century, at least in the northern half of the country, where they represented up to a third of equid production. This indicates how important this animal was for the Romans; it must have been used for many tasks, primarily that of ensuring transport within an empire whose economy had been globalized from the north of Great Britain to the shores of the Black Sea, over more than

5,000 kilometers. In particular, it played a strategic role in provisioning the armies scattered over that huge space. In this, our genetic results echoed the texts of authors such as Pliny the Elder, who reports that the use of donkeys was reserved solely for breeding mules, or Columella, who listed all the circumstances in which it was appropriate to resort to mules in the first half of the first century, because they surpassed the performance of horses. The omnipresence of mules in Roman era sites that we studied reminded us of an already obvious truth as seen in the mosaics and bas-reliefs of the period: the animal was used for just about everything, turning millstones, pulling military carts or *vallus*, the Gaulois ancestor of the harvester. A do-it-all animal, in short, as indispensable in the daily lives of people as in the life of the empire.

Our data revealed an exception, located at a site in Boinville-en-Woëvre. In a countryside located not far from the city of Metz, at barely twenty kilometers from the Belgian and Luxembourg borders, there is a Roman villa that can be dated from between the third and sixth centuries CE. The remains analyzed come from the *pars rustica* of the building, the sector that was reserved for agricultural work. In that place we didn't find (as we did almost everywhere else at the time) remains representing the expected proportion of one-third mules and two-thirds horses, but almost entirely those of donkeys. There were males and females, enough to maintain husbandry there. But for Lepetz the results were explosive, because the remains he had sent to me were those of singularly large animals: giants measuring 1.5 meters or more to the shoulder, which for donkeys, whose average size is more around 1.2 meters, is exceptional. Lepetz would never have imagined that they could be donkeys. And there were more surprises to come, since the donkeys we had analyzed were genetically related: six of them came from the same family, and two from another. They hadn't found themselves in Boinville-en-Woëvre by chance, gathered by Roman breeders from four corners of the neighboring countryside; instead they had been raised on-site, in a family. To complete the tableau, one of these animals was none other than the fruit of a strictly consanguine crossing between a brother and his sister. Such a union occurs only rarely in nature, and so it, too, was likely the result of a choice made by the breeders.

Around fifteen centuries ago, the breeders who worked at the rich Roman villa of Boinville-en-Woëvre produced families of donkeys with exceptional physical characteristics, and presumably sought to maintain and even to improve them, to the point that they likely encouraged notably consanguine crossings to achieve their ends. The genetic data indicated that the site was used for breeding donkeys of a special type, but also informed us about the methods used. The texts of the period, including those of Pliny, Columella, and also Varro, tell of the use of gigantic donkeys in the Roman mule-breeding industry and the existence of what appears to be true organized breeding centers. The circle was closed: the breeders of Boinville-en-Woëvre raised donkeys that were to be used primarily for producing mules. In addition, since we had identified no male mule, no female mule, and only very rare mares on-site, the breeders must have mainly rented out the services of their giant jack donkeys to surrounding farms. There were two possibilities: either the owners of mares brought them there for a spring fling, or in the right season the breeders went around to neighboring farms with their donkey, which then became an itinerant genitor. The advantage, in both cases, was that the mule would have been born directly on the owner's property.

There was a last and equally important piece of information in the genetic profile of the donkey remains from Boinville-en-

Woëvre. If most of the donkeys have ultimately proven to be very close to those that are found today in Western Europe, they all showed a particularity that is not found in donkeys from other periods: a slightly more pronounced genetic proximity to groups of wild donkeys. Some even showed genetic affinities with donkeys that live today predominately in West Africa, from the coasts of Senegal and Mauritania as far as Mali. This was a sign that perhaps the Roman breeders of Boinville were able to benefit from the immense expanse of the empire and the enormous natural resources it offered beyond the seas, particularly access to the large donkeys that would serve their interests so well. The idea wasn't so outlandish; after all, isn't that what George Washington had done in requesting Andalusian donkeys? Even if those donkeys, according to the legend, came from Egypt in the seventh century BCE and not from West Africa, weren't they also, at a height of 1.5 meters or more to the shoulder, true giants?

The Animal of Kings

I wouldn't do complete justice to the hybrids we're looking at here if I didn't mention another of their uses beyond that of beast of burden with the so maligned "mule head": prestige. Hybrids were the fruit of a union against nature and owed nothing to chance, but to the know-how that only a few knew how to exploit. And in the most distant of times, when that rare, still uncertain know-how was the precious property of only a few, the hybrid was therefore an animal that only the richest could afford. In those conditions, should we really be surprised to see King David of the Old Testament appear on muleback? At the time, it wasn't rare in the East for dignitaries to ride a mule or other equids from hybrid unions—such as the ones that gave birth to the animal today called the Kunga, made famous in cuneiform inscriptions of the second half of the third millennium BCE and designated by the term ANŠE.BARxAN: the equid of kings. According to specialists, the Kunga is believed to be the mysterious equid appearing in Sumerian mosaics on the Standard of Ur, that wooden box decorated with mother-of-pearl, limestone, and lapis-lazuli found in the 1920s south of Baghdad close to 4,500 years after it was created.

We also now know with certainty that the Kunga, which sold at exorbitant prices—six times more than the donkey—was the offspring of a female donkey and a Syrian onager, ever since DNA traces from skeletons over 4,000 years old found at Umm el-Marra have spoken. There is no longer any doubt that the end of the third millennium BCE was a period of experimentation without taboos concerning the art of breeding and raising equids. Once again, let's remember that the ancestor of our modern horses was at the time germinating in the steppe of the North Caucasus. A bit farther south, on the other side of that same mountain chain at the doors of Anatolia and in Mesopotamia, other peoples also thought of intervening in the reproduction of equids by crossing wild males of the local species of onagers with domesticated females of a new species that a few centuries earlier had left the Nile valley and the Horn of Africa where it was born: the donkey. Our genomic work has enabled us to solve certain mysteries concerning the domestication of that distant cousin of the horse—the animal in our spotlight, whose story we return to now.

7

The Horse of the East

The Arabian: A Legendary Horse

With its concave forehead, long, arched neck carrying a proud head, and flamboyant tail set very high on its rump, the Arabian stands out among all horses. The magnificent stallion Marwan Al Shaqab is the current reigning champion of the World Arabian Horse Championship, which brings together the global elite of these horses to crown the undisputed winner every year at the Porte de Versailles. Its owners wouldn't be separated from it for anything in the world, not even for the hefty sum of $20 million that has been offered. The stallion incarnates the absolute beauty of a breed of horses whose origin is lost to time. According to the most popular legend, their origin goes back to the prophet Mohammed, who, after a long journey through the desert, released his horses when he saw an oasis so they could quench their thirst. They didn't have to be asked twice and went off at a gallop, but they hadn't reached the water before their master called them back. Only five mares stopped running and turned around; Mohammed then knew which horses he could count on, come what may. They remained throughout history "the five," *al Khamsa*, faithful among the faithful, those who would give birth to the horses of the Bedouins of the desert. These horses were a particularly resilient breed among which only the most perfect would be allowed to reproduce, and whose most precious mares had a place in the family's tent to shelter them from thieves. Other legends place the origins of the Arabian in even more distant times, back to Allah, who gave this creature a body from the winds of the south: "I have made thee as no other. All the treasures of the earth lie between thy eyes. Thy shalt carry my friends upon thy back. Thy saddle shall be the seat of prayers to me. And thou shalt fly without wings, and conquer without sword; oh horse."

Recent archeological discoveries at Al Maqar, 800 kilometers southwest of Riyadh, in Saudi Arabia, also suggest origins dating much earlier than Islam. There is one find of particular interest: a carved stone 86 centimeters long and weighing more than 135 kilos whose profile strangely resembles that of a horse. Beyond its size, the piece is notable because of a very obvious vertical protrusion at the base of the neck that appears to be a rope, or an element of a harness that could have been used to hold the animal. But the stone also stands out due to its age: over 6,600 years old, maybe even up to 7,300, according to the radio-carbon dating carried out on the bone fragments found next to the sculpture. This is more than a thousand years earlier than the Botai horse is believed to exist, and at least 2,500 years before the advent of the DOM2 in the Pontic steppe. Could it be possible that the Bedouins were the first to have domesticated the horse? After all, until around 6,000 years ago, the Arabian Peninsula wasn't the great desert we know today but instead a more humid region—in fact, a lush savanna. So why couldn't it have had horses? Without a doubt, the discovery was sensational.

With this discovery, the debate was launched. The huge object hadn't been found in a supervised archeological excavation:

the man who discovered it had unearthed it when he was building a cistern, and instead of leaving it where it was, put it in his
car with more than 300 other objects found nearby to take them
to Riyadh, to the Saudi commission in charge of tourism and
antiquities. So there was no way to tell if it was the same age as
the bone fragments found near it. Before this discovery, the oldest carved rocks found in this region depicted mounted horses
or horses harnessed to carts and dated back to around the beginning of the second millennium BCE; their age corresponds
to the expansion of the DOM2 horses. In addition, none of the
engravings (petroglyphs, as archeologists would refer to them)
show incontestable attributes of the Arabian horse—the concave forehead, arched neck, and high tail—before the ninth
century BCE. While the profile of the sculpted animal found in
Al-Magar appears to be a horse, it could also very well be another equid such as an onager or a wild donkey. And it might
be jumping the gun to assert that the band sculpted in relief at
the neckline represented elements of a harness.

There are other troubling elements. Neo-Assyrian royal an-
nals mention many diplomatic gifts from the Arabian Peninsula
starting in the twelfth century BCE; we find donkeys and drom-
edaries, but never horses. The "Procession of Tribute Bearers"
frieze constructed at the end of the sixth century BCE in Perse-
polis, Iran, which appears on the Apadana Palace walls and
shows delegations from Armenia, Cappadocia, and elsewhere
bringing gifts to the Achaemenid king Darius I, depicts camels
and not horses that seem to have made the journey. Furthermore, the word "horse" is completely absent from the corpus
of texts written in Minaic, a semitic language dating from the
eighth to the first century BCE. Researchers found it only once
in examining 2,500 inscriptions of the Qatabanic textual corpus,
written in another semitic language dating from the seventh

century BCE to the second century CE. And it is only after the first century CE that its use seems to have spread into the textual corpus of another neighboring semitic language called the Sabaic language. That appears to be very late to make the association with a sculpture that presumably was created 4,000 years earlier. One can justifiably ask: Where are the traces of horse skeletons that should be abundant if they had been domesticated by that Neolithic civilization? Even though King Abdallah of Saudi Arabia declared that the discovery of Al-Magar should be expediently published because it proved that "the Arabian Peninsula was the first to have domesticated horses," the fragments recovered were quite sparce, and understandably far from convincing.

The Origins and Expansion of the Arabian Horse

What does DNA have to say about all this? It is, to tell the truth, much more unanimous. The Arabian horse belongs to the genetic lineage of DOM2: the same horse that was domesticated in the lower Don and Volga valley 4,200 years ago along with the horses associated with Sintashta carts at the beginning of the Bronze Age; the horses of the Scythian nomads of the Iron Age; and the Celtic, Viking, and Byzantine horses whose DNA was also sequenced in my laboratory. All of that is beyond any doubt. If we continue to think that the Al-Magar sculpture reveals the existence of a domestic horse around 7,000 years ago, then we must accept the evidence: that horse bequeathed no direct descendant to other historical horses, including Arabian horses. However, even if the DOM2 horses are all genetic siblings in a way, it nevertheless remains true that some resemble each other more than others. By exploiting these relative genetic proximities we can deduce which ones gave birth to others, once the process of domestication had been launched. We have been able to show that contemporary Icelandic horses descended from Viking horses, or that the horses of the Xiongnu nomads, the same ones who harassed the Middle Kingdom of China in the centuries around the beginning of our era, derived from the horses used by the Pazyryk Scythians from Berel', a site located in Kazakh Altai. In France, Celtic horses were closely related to Roman horses. This type of exercise reveals something important about the origin of today's Arabian horses: that it was necessary to go back at least to the Sassanid Persian horses of the fourth and fifth centuries CE to find their genetic source.

Genetically, the Arabian horse was already germinating before the rise of Islam, in the heart of the empire that established Iran as a superpower in late antiquity. Without having found fossils in Saudi Arabia whose DNA was usable, we can't say for the time being whether the horse was born in the heart of the Arabian Peninsula before going into Persia or if the opposite is true. We are working hard to answer that question, but without too much risk we can confidently assert one thing: the Arabian profoundly changed the history of modern horse breeding. Many modern breeds of horses—with a few rare exceptions, like Shetland Ponies or those of the moorlands of Dartmoor in the British Isles, or the horses of Iceland—owe something to it. Take the Indian Marwari, for example, whose courage in battle is legendary. It's commonly said that there are only three ways a Marwari will leave a battlefield: victorious, all four legs up in the air, or bringing back wounded soldiers. This horse, with ears shaped like a crescent moon, almost disappeared under the English colonial occupation; fortunately it was saved, and today continues to be the pride of the inhabitants of Rajasthan, its native land. However, although native to this province in northwestern

India, the horse inherited more than a third of its genome from Arabian ancestors. Those horses were surely from among the ranks of the tens of thousands of horses that were taken across the Indian Ocean in the holds of ships from the coasts of Yemen beginning in the twelfth century, contributing to the establishment of the authority of the Turkish sultanate of Delhi and soon expanding to most of the Indian subcontinent.

The Marwari is only one example. Our genomic data show that the horses that populated the European continent in late antiquity, and that could still be found along the coasts of the Baltic in the ninth century CE, would soon undergo many changes connected with Arab-Muslim expansion. The horses buried in Croatia between the seventh and ninth centuries CE would begin to show a genetic affinity with Eastern horses, which until then was unknown in that region of the world. The same is true for the military horses found around the ancient port of Byzantium, whose activity culminated between the fifth and eleventh centuries. The influence of the Arabian horse was also felt in central Asia and even in Mongolia, and we have found its genetic trace in horses buried in Mongolia from the thirteenth century, when the empire of Genghis Khan was at its height. We are not the only ones to paint such a portrait; Barbara Wallner of the University of Veterinary Medicine Vienna has also demonstrated the impact the horses of the East have had on modern horse breeding.

The Unparalleled Success of Arabian Stallions

Wallner's focus isn't in the entire genome but one of its chromosomes in particular: the Y chromosome, and more precisely that part of the chromosome that never combines with the X chromosome. The part that fascinated her is often called the "male specific [region] of the Y chromosome" or the MSY region, since it carries the gene that drives the development of the male genital apparatus; those who carry it are always male. For a geneticist like Wallner, this region is very useful because it enables one to follow the evolution and domestication of the horse from father to son, along paternal lineages: "Show me your Y chromosome and I'll tell you who your father was," in a sense. If Arabian stallions were as successful as we've suggested, presumably being used to cross with local mares just about everywhere around the world, Wallner would be capable of finding that in her genetic data because the foals born of such unions would tend to carry the same version of the MSY region as that carried by the genitor stallions: a principle powerful in its simplicity.

We have known for around twenty years that the domestic horses living today are almost exceptions in the animal kingdom; they are domestic animals for which the genetic diversity of the MSY region is the most limited. For a long time we even believed that all the domestic stallions in the world carried a single and same version of the Y chromosome. Sequencing techniques have improved a great deal since then, and we have been able to find that a hidden diversity existed, but that it remained exceptionally limited. Though the details have changed, the tendency has still remained the same; with very few exceptions, most contemporary horse breeds carry versions of the MSY region that are not only very close, but above all that derive either directly from Arabian horses or from English Thoroughbreds whose paternal origins also go back to the doors of the East, or from the Barbs of North Africa and their Ibero-Andalusian descendants. By using the rare mutations that differentiate these close versions and by applying the principle of molecular clocks that provide the rhythm at which the mutations appear over time, Wallner has even been able to estimate that this group had its roots between the sixth and the fourteenth centuries CE; this tells us when the stallion ancestral to that lineage last existed, before mutations transformed the Y chromosome it carried into the different versions now found in its descendants. That period perfectly coincides with the moment when we, too, observed the surge of an important change in the genetic composition of domestic horses preserved in the archeological register of Europe, dating from the ninth century. It is very likely that the stallions that came from the East—and the Arabian horse in particular—made a huge impression on medieval breeders, so much so that they had a massive (and lasting) influence on the future of European equid husbandry. We know that this influence wasn't uniform but instead multifaceted, because Arabian, Barb, and Andalusian horses do not uniformly carry the same Y chromosome as those that ended up giving rise to the English Thoroughbred. Although they both came from the East, Arabian horses and the ancestors of English Thoroughbreds essentially show marked differences in all of their 31 pairs of chromosomes not involved in determining sex. Evidently, the breeders also tapped other Eastern sources over the course of history as well.

A Quest for Elegance, Grace, and Endurance

What those breeders were looking for largely remains a mystery, and it is just beginning to be solved. In particular, we have endeavored to uncover the regions of the genome where the descendants of the Sassanid Persian horses (third–seventh centuries CE) and the Byzantine horses (fourth–fifteenth centuries CE) differed most from the horses that populated Europe and Asia before them. This work has enabled us to identify a group of genes that share a key role during embryonic development; they are responsible for organizing and assembling the bones of the skeleton, notably types and numbers of different vertebrae. When we realize that Arabian horses often have one less lumbar vertebra and two fewer ribs than other horses, it seems reasonable to imagine that it was the search for a new aesthetic profile that encouraged breeders to introduce Arabian blood into their stock beyond the East.

The Arabian horse is known as much for its characteristic appearance and natural elegance as it is for its exceptional endurance. The hardiest among them face off in races over 160 kilometers long, during which they complete four marathons interspersed with rest periods, and are able to carry their rider while maintaining average speeds of 20 kilometers per hour. Such phenomenal prowess is only possible with training, of course, but it is also their natural abilities that equip them ideally for that type of race, which makes them the perfect subject for study for geneticists who want to solve the physiological mysteries of endurance. Éric Barrey, research director at the Institut national de recherches pour l'agriculture, l'alimentation et l'environnement (the INRAE), is one of them. I had the pleasure of working with him while I was attempting to sequence the genome of ancient Arabian horses, and he's the one who convinced me to turn away from the archeological collections I was using and to look at the collections of the Musée de l'Armée, in Paris. One of the museum's most famous guests had just been restored: Le Vizir, a silver-grey horse from the early 1800s ridden by Napoleon in the Battle of Jena, such a favorite that he accompanied the emperor into exile on the islands of Elba and Saint Helena. Le Vizir was not large, measuring barely 1.35 meters; we could see the crowned N of Napoleon, the brand of the imperial stables, on its left thigh. Barrey and I had never been so close to history as we were that day when we obtained an official permit to remove a few of the animal's hairs from the hollow of its pastern, just above the hoof toward the back, in a place where no one would notice what we'd done. We were ultimately able to characterize its genome, but that is another story. Let's go back to the biology of endurance.

The Biology of Endurance

It takes a complete set of biological conditions for a horse to be able to continue running for long periods of time, starting with its ability to produce the energy the muscles need to function. For that, the organism must pull from its reserves—fats (triglycerides) and sugars (glycogen)—and convert them into a cellular energy fuel, which it does with the help of two important types of metabolic reactions: those that consume oxygen (aerobic metabolism) and those that don't (anaerobic metabolism). If all the cells that make up muscle fibers stock reserves of glycogen, those fibers are not all the same; some of them, called type I, are relatively slow to contract, while others, type II, are much more rapid. Among the type II fibers are those called IIA, that also stock reserves of fat and are, like type I fibers, capable of drawing energy from them by consuming oxygen. The others, type IIB, don't have any fat, and are specialized in the anaerobic metabolism of sugars.

At a walking pace, it is mostly the slow type I fibers that are mobilized. The organism draws its energy through paths that consume oxygen. When it accelerates, more rapid muscle contractions are necessary; at a trot, it is the rapid type IIA fibers that essentially ensure that task, and aerobic metabolism is still sufficient to supply energy from the fats and glycogen present. But at a gallop, and as soon as the speed really accelerates, anaerobic metabolism kicks in to provide the organism the boost of energy it needs: the rapid type IIB fibers are mobilized. If this path has the advantage of supplying the fibers the energy they need to sustain the rhythm of rapid contractions, it has the enormous disadvantage of acidifying cellular pH, because it leads to the manufacture of lactic acid, the molecule that engenders muscle fatigue and soreness after exercising. In the long term, during sustained exercise all energy is obviously worth having, and the formation of lactic acid is inevitable—this is the case even at less intense paces, especially when they are prolonged.

It happens that the muscles of Arabian endurance horses contain greater proportions of type I and type IIA fibers than do the muscles of sprinters. They are naturally able to maintain trotting paces thanks to a metabolism that consumes oxygen without the risk of muscle fatigue. Exercise and training can promote a balance of fibers, as well as an increase in the mitochondrial density of muscle (those mini cell factories in which oxygen is consumed), and the vascular tree of fiber clusters. All of this contributes to improving the oxygen supply to the muscle, and its ability to maintain the effort of contraction without tiring. A notable difference: in Arabian horses, the balance is naturally favorable, because it is provided at birth.

But there is more. Their type I fibers also naturally produce a large quantity of the protein SLC26A1, which serves to pump and redirect to the other muscle fibers the lactic acid that will inevitably be formed, sooner or later, in the type IIB fibers. This presents a double advantage. First of all, if lactic acid is no longer accumulating in the type IIB fibers, muscle fatigue will be limited or will take much longer to settle in. But since type I fibers, in addition to SLC26A1, contain other proteins that are able to produce their energy fuel from lactic acid and oxygen, those cells, following the work of SLC26A1, recover a surplus of energy and can contribute longer to the effort. This energy renewal isn't possible in type IIB fibers, in which the proteins necessary for that conversion are lacking. Upon arrival at a finish line, the muscle is overall less tired and recovers more energy from the same quantity of reserves; the horse is able to draw from its reserves during sustained effort. In a word, it has more endurance.

A mutation of a single letter in the gene coding for the protein SLC26A1 is particularly frequent in Arabian horses, champions of speed races, and for a good reason: it leads to an increased production of the protein, which in part explains the exceptional ability of those horses to push the limits of their effort. If you change an A to a G at that precise location on the genome of an Arabian horse without touching the 3 billion or so other letters, its chances of emerging victorious from a race would increase quite significantly. It is the carriers of this natural mutation that some breeders of Arabian horses built for racing have, unbeknownst to them, selected, contributing to making this breed the legend it has become.

Obviously, not everything can be summed up in that single genetic change; many other differences also intervene, the very ones that people like Barrey aim to discover. There is, for example, another mutation of a single letter that affects a gene whose product enables muscle fibers to burn their reserves of fats in the presence of oxygen, and thereby obtain the energy fuel necessary for the effort. It is called *ACOX1*, and horses that carry a T at a precise location on this gene will on average ensure their owner more wins in speed races than those who carry a G, who are more suited for endurance races and better built to survive in difficult environmental conditions such as those of the Arabian deserts. The gene *ACTN3* and its product also play a key role in the physical phenomenon of rapid muscle fiber contraction: it is again a single mutation—a T in the place of a C—that intervenes, generally leading to better natural endurance.

The list of genetic factors explaining the qualities of Arabian horses continues to grow longer, as the studies comparing their performance at a gallop or in endurance efforts, or in measuring their physiological state before and after effort, multiply. Some of those factors favor the metabolism of fats during prolonged effort. This does not necessarily involve single mutations; many genetic modifications can be involved, the whole contributing to modulate the genes expressed within the muscle fibers, and the biology they are capable of putting to work. Others act directly on certain nerve cells in the brain; they probably participate in the complex processes that forge the competitiveness of high-level athletes, who well know that races are won as much with the head as with the legs. Taken separately, the contribution of each of these genetic factors in the performance of an animal is nevertheless minimal; the contribution remains extremely difficult to detect even using the most effective tools, unless thousands of subjects or more are included in our studies. Spectacular progress has been made; nevertheless, though some of the biological secrets of Arabian horses have been revealed, others remain, both due to their complexity and their subtlety. The legend keeps part of its mystery, even today.

The Genetic Flipside of the Coin

The history of equid breeding, specifically that of Arabian horses, resulted in very beautiful animals that are among the most resilient on the planet. That's the glorious side of their history. However, there is a flipside. Arabian horses are prone to genetic syndromes, and sometimes in proportions that are truly alarming. Some of these syndromes are not unique to Arabian horses, but also strike other breeds. But for those unique Arabians, the outcomes are most often fatal. For example, there is the severe combined immunodeficiency syndrome, better known by its English acronym SCID, which can appear in a newborn even though its two parents show no sign of the condition. At birth the newborn will appear completely normal, but its body will eventually prove incapable of producing two types of white cells, the B and T lymphocytes, that are crucial for the immune system to respond to viral and bacterial infections. The result: the foal will be healthy as long as the reserves of antibodies that its mother transmitted to it during pregnancy aren't exhausted (this is called passive immunity), or for several months. But afterward, it will not be able to develop its own lymphocytes and antibodies, and will then be at the mercy of the slightest opportunistic germ. On average it will not live beyond the next three months, even if it is given the most intensive veterinary care.

Today, we know the mutation responsible for this condition. It affects only one gene, the *DNA-PK* gene, which codes for a crucial enzyme in the maintenance of the entire genome. Among other functions, the product of this gene, the enzyme, intervenes at a key stage in the reshuffling of the genome that accompanies the formation of new types of antibodies by the lymphocytes. Therefore a deficiency of that enzyme unsurprisingly leads to problems in the development of those cells and a loss of the ability to produce one's own antibodies. The genetic mutation consists of a deletion that shortens the size of the gene by only five letters. The problem is that the place where this deletion is located corresponds to a zone that codes for a crucial module in the activity of this gene's product. By losing five letters in that place, the text carrying the genetic information becomes displaced, which is particularly serious when one knows that the genetic code functions in multiples of three letters and that five is not a multiple of three. The enzyme manufactured from that mutation will have nothing in common with the normal enzyme; it will be less stable and above all nonfunctional. In individuals carrying a single version of the defective gene (such as the one transmitted by their father), the second available version (the one transmitted by the mother) is intact and can be enough to produce the normal enzyme; the individual will not develop the condition and the mutation is called recessive. But in those carrying two defective versions (transmitted by both parents, both healthy carriers), there can be no cure and the outcome will always be fatal.

An Optimistic but Lucid View

Genetic tests have been developed to help breeders detect healthy carriers of the mutation affecting the *DNA-PK* gene, so that they can avoid crossing them with other healthy carriers. Without that help, one newborn out of four would certainly be stricken with the condition; one out of two would be a healthy carrier, and only one out of four would not carry the mutation responsible for the condition. The increase in such testing has revealed the extent of the problem, since in just the population of Arabian horses bred in the United States, close to one animal out of ten carries the mutation (some estimates have even suggested that one animal out of four could carry the deficient version of the gene). This is proof—if any is needed—that genetics doesn't just inform us about the historical past of these animals; it can help us here and now to prevent unnecessary suffering and improve their health in the future.

There is reason for optimism in this regard, because other genetic tests have been developed to detect mutations responsible for other genetic syndromes affecting Arabian horses. Lavender foal syndrome (LFS) is one example. The recessive mutation involves the loss of a single letter in the *MYO5A* gene, and results in complete muscular weakness; the afflicted animals have compromised limbs, necks, and backs; they can't stand up, and are prey to rather traumatizing epileptic episodes. The foals must be euthanized because there is no possibility that they will be cured. Nevertheless, the mutation is very frequent among Arabian horses from Egypt; more than one animal out of ten is affected, and stallions, particularly prized by breeders, are not spared. Other just as tragic syndromes threaten Arabian horses, such as cerebellar atrophy, which affects the cerebellum and leads to serious balance problems that increase the risk of accidents. Fortunately here, too, breeders can again benefit from genetic testing.

However, it would be a mistake to believe that the availability and increase in genetic testing will solve everything, like waving a magic wand that will eradicate deficient mutations from the planet. The example of HYPP syndrome is proof of this, and this particular genetic disease (which doesn't affect Arabian horses, instead afflicting the American Quarter Horse) can be useful in elucidating one of the fundamental limits of genetic screening. HYPP, or hyperkalemic periodic paralysis syndrome, is the first genetic disease in horses to have been characterized. It was in 1992, when the genetic mutation had just been identified. Unlike SCID or LFS, there is no healthy carrier of this mutation; the mutation is called codominant. The symptoms are more serious in individuals who possess two deficient versions (called homozygotes) instead of just one, so that two-thirds of them will not live past their sixth year. Despite the

availability of a genetic test and a ban on including the homo-
zygote carriers of the deficiency mutation in the official Quarter
Horse register, the mutation has spread widely since 1992.
Thirty years ago its frequency was 20 percent; today, it surpasses
50 percent. This is because breeders have found that diets low
in potassium can significantly mask or suppress symptoms of
the disease in carrier animals while they are of prime competi-
tor age. It also happens that the animals whose appearance conforms best to the aesthetic preferences of judges in competition
are more often carriers of the mutation. Carriers benefit from a
boost in competition and will earn more awards, which means
they will most often be chosen for reproduction. The vicious
cycle continues, because the mutation they carry can quite
clearly be an advantage to winning, but will also be propagated
willfully within the population. This is made all the more possible since the new diet limits the risks that afflicted individuals
will become sick. The irony of the story is this: the genetic test
initially developed to impede the propagation of the mutation
in American Quarter horses can today serve the interests of un-
scrupulous breeders who want to identify it among their animals in advance—solely for the purpose of singling out those
who will have the greatest chances of success in competition.

Even if the HYPP mutation doesn't affect Arabian horses, it
is clear that we are touching on a fundamental principle in
breeding: its history as well as its current development can lead
to situations that are not always foreseeable. If it is then completely reasonable to imagine that genetics will have something
to say in the future of the Arabian horse, it is just as reasonable
to think that technology will not necessarily be adopted as the
only justice of the peace. In this regard, it is appropriate to keep
in mind that all mutations responsible for serious illness are not
known; far from it. Among Arabian horses, for example, the

genetic deficiency that induces the juvenile epilepsy syndrome remains a complete mystery to this day. But beyond this simple admission of ignorance, there exists in populations a large number of harmful mutations that are not as clearly manifested through such harmful effects on health as those we have just looked at. Even if their individual effects are weak, collectively, they can ultimately diminish an animal's life expectancy or impact its fertility. To ensure that only a handful of mutations are absent from the genome of an individual is not enough to guarantee its future health and well-being. Being aware of its overall genetic burden—that is, the sum of the harmful mutations present in its genome—might provide information on this subject and have a useful influence on the choices of breeders.

Genetics as a Burden, and Its History

If genetic testing is seductive in principle, it also faces a sizeable practical difficulty: most often we don't know in advance if a mutation is harmful or not. Taken in isolation, a mutation's effects are assumed to be weak; most of the time they will be too subtle to be measured individually. In addition, since they might affect all biological phenomena, they are difficult to detect; we would scarcely know where to begin, or what to measure. To overcome this difficulty we have decided to look to evolution, because it has left some precious clues for us to use. If mutations are truly harmful, then they should not be very frequent in the living world. By definition, the individuals that carry them have fewer chances of surviving and reproducing, and therefore of propagating them from generation to generation. It then becomes possible to estimate the risk that a mutation carried by a horse is harmful without directly measuring its effects on the animal; a mutation will be all the more harmful if it affects a place on the genome where all other mammals carry the same other letter instead. With a few mathematical tricks, one need only identify the mutations present in the genome of an individual and calculate their risks the way evolution enables us to predict, to measure the individual's genetic burden. Ingenious, right?

This is the simple principle we decided to apply to horses living today, as well as to those that lived in the past. Our findings were conclusive: for close to 4,000 years, breeders managed to keep the genetic burden of their horses just about constant. The genetic health and well-being of their animals have remained the same and have not gotten worse. But starting around the eighteenth century the situation changed dramatically, since the portion of harmful mutations present in each genome increased considerably. We were able to determine that in barely two centuries, in the part of the genes coding for proteins, the portion of harmful mutations increased by around 4 percent, and in the intergenic regions by around 1 percent. It's important to understand that from a genetic point of view, the heftiest genetic toll the horses have had to pay in the course of their history doesn't concern the moment of their domestication, nor the close to 4,000 years that followed, but the last two centuries. In other words, the period when the practices of modern breeding, in closed groups, have enabled the emergence of the biological breeds we know today, but whose collateral effect has been to concentrate more harmful mutations in the genomes of each individual. The hypothesis called the "cost of domestication," which assumed that domestication could have led to this phenomenon from the very beginning, had to be entirely revisited. It is not domestication as such that is to blame for these more harmful genetic mutations in horses, but modernity.

There are breeds that have paid a higher price than others—traditional draft breeds whose use considerably declined during the twentieth century, for example. But the Arabian horse seems to have suffered less than those horses, and we can be happy about that, especially since entire cradles of genetic diversity are beginning to be discovered among the Arabian horses of Iran, Bahrain, Syria, and Tunisia. These genetic reservoirs might well be mobilized to be introduced to today's Arabian horses and enable us to envision a positive future for those breeds.

The Khomiin Taal Plateau in Mongolia, May 2014. One of the rangers from the neighboring Seer reserve, where the TAKH association began reintroducing Przewalski's horses in the mid-2000s, helps us round up domesticated horses. The objective: to study their DNA and compare it to that of horses that lived in this region in the time of Genghis Khan.

The Botai archeological site, evidence of the existence of a sedentary people who bred horses in the heart of the Kazakh steppe close to 5,500 years ago.

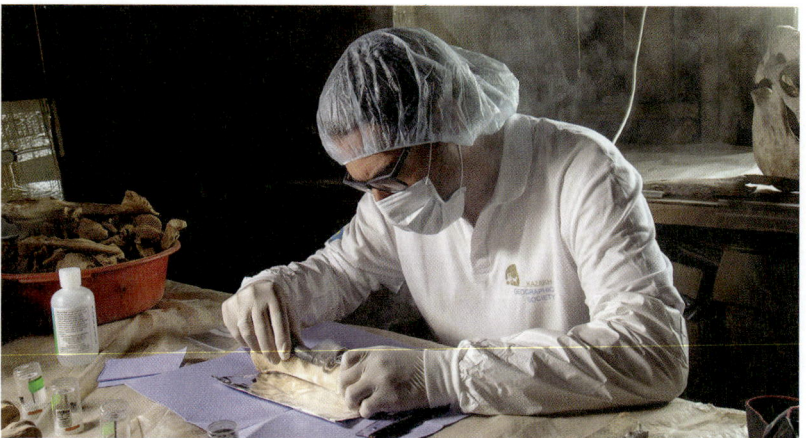

The author sampling horse bones found on the Botai archeological site for later genetic analysis in clean lab facilities (© aAron Munson, Handful of Films).

A portion of the collections in the archeology laboratory at Samara State University, in which the author, guided by Natalia Roslyakova and Pavel Kuznetsov, borrowed a hundred or so archeological vestiges from western Russia going back to a pivotal transition in the history of the domestication of horses: from the fourth to the second millenium BCE.

The Pavlodar region, August 2016. A Kazakh woman milks a mare, performing a task that the people who lived some 50 kilometers from there, in Botai, undertook more than 5,000 years ago.

Early July 2019. Breeders in Inner Mongolia show me how they round up their horses. In the distance are wind turbines, signs of a new era in which modernity collides with traditions that go back many thousands of years.

A Soviet kolkhoz located not far from Kislovodsk, converted into a warehouse, where archeological remains of one of the direct ancestors of modern domestic horses are stored.

Przewalski's horses, which the TAKH association reintroduced into their natural habitat in the Seer reserve in western Mongolia. The reserve, and its breathtaking landscapes, contained some fifty horses when I visited it for the first time in May 2014; today, it has more than one hundred.

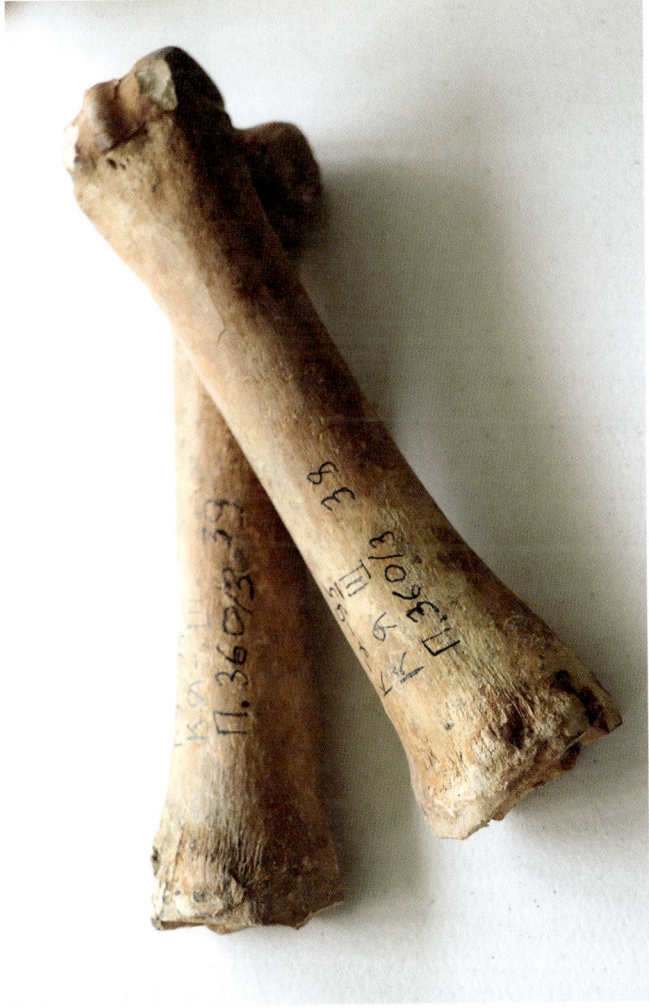

Some bone remains from one of the 4,000 horses analyzed as part of the ERC Pegasus project. Only a few hundred milligrams of material will be reduced to powder to assess whether ancient DNA of sufficient quality is preserved and determine the complete genome sequence of an animal that was buried nearly 1,500 years ago.

Descent into the depths of the sinkhole of Igue du Gral, near the Pech Merle cave. This natural trap, in the Quercy bedrock, was fatal to many prehistoric animals, whose remains, dating back more than 15,000 years, still provide DNA.

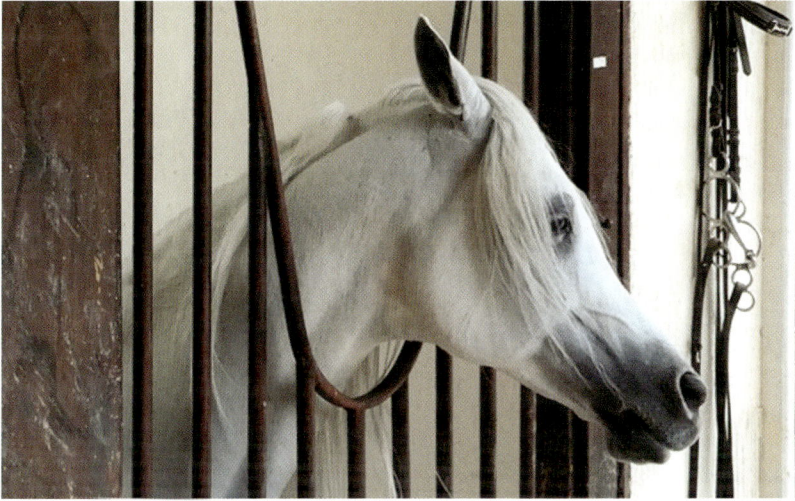

An Arabian horse with its characteristic forehead, patiently waiting for us to draw its blood in a stable on the outskirts of Riyadh, Saudi Arabia, February 2019.

Dr. Yvette Running Horse Collin with her people's horses in He'Sapa, the Black Hills of South Dakota. My collaboration with Yvette has gone well beyond our common passion for horses. She has contributed to rebuilding connections between two worlds that history has often put in opposition, but that nonetheless have so much to share. Photograph by Lyla June Johnston.

A Yakutian burial site from the first half of the eighteenth century discovered in Alaas Ebe (Yakutia). The horse and its tack have been perfectly preserved by the extreme cold of northeastern Siberia. (Photograph courtesy of Expedition MAFSO 2002 [Mission Archéologique Française en Sibérie Orientale], directed by Professor Éric Crubézy. © MAFSO 2002)

Toward the end of the nineteenth century, photography provided the opportunity to break down the movements of animals beyond what our eyes were able to see. The German photographer Ottomar Anschütz was one of the pioneers in this realm, and here shows dressage work with a Hanoverian Horse, a breed refined over time through crossbreeding with English Thoroughbreds (Lissa, Prussia, 1884, private collection). At the same time, on the French side, Étienne-Jules Marey was attempting to understand the locomotion of what he called the "animal mechanism" through chronophotography. His work revolutionized our understanding of the gallop as a four-beat pace, during which each foot hits the ground separately.

Fields at the Doña Sofía station, Argentina, December 2021. Lolo is training a clone to trot while wielding a polo mallet, in the hope that one day, the horse will become a polo champion.

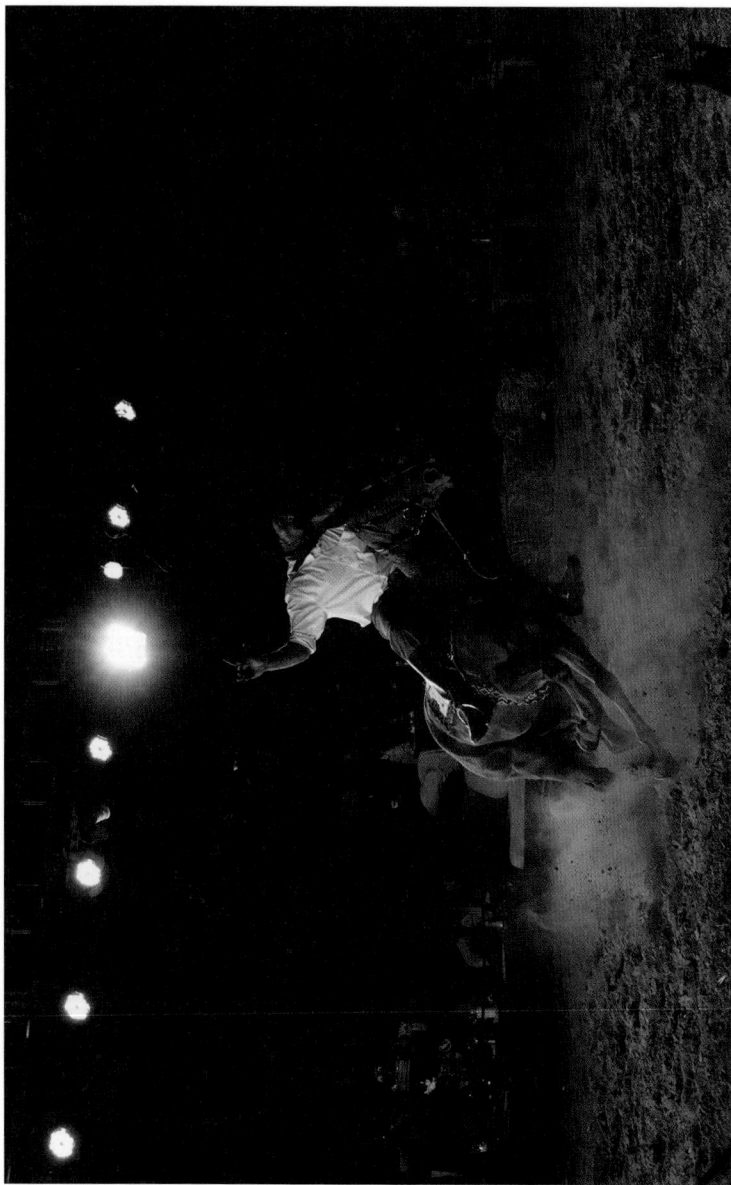

Rapid turning demonstration during an auction of polo ponies, in the outskirts of Buenos Aires, December 2018.

On the Kazakhstan steppe, on a rock face, artists engraved for eternity the profile of horses that lived around them many thousands of years ago. The image shows two types of horses that coexisted: possibly the DOM2 lineage, the mother of all modern domestic horses, and the Botai lineage, the origin of Przewalski's horses. Will we ever really know? (Photograph by Victor Merz.)

The Duruthy Horse, discovered in a rock shelter in the French Landes and preserved at the Arthous Abbey museum. The figurine, carved out of ivory around 17,000 years ago, shows an attentive animal with its ears cocked and nostrils flared as if breathing. Although this animal is not related to our domestic horses, we can appreciate the keen observation of the artist, who was able to understand and fix for eternity the very essence of this horse that no longer exists today.

8

The Horse in the Middle Ages

The Horse, the Church, Texts, and Images

"And another, a red horse, went out; and to him who sat on it, it was granted to take peace from Earth, and that men would slay one another; and a great sword was given to him."

—*REVELATION 6:3–4 NASB*

Red—actually, chestnut. That was the color of the horse ridden by the second of the four horsemen of the Apocalypse, the one through whom arrived the first of the great plagues: war. It was on the back of a black horse that the next horseman sowed famine and brought another equally terrible plague to humanity. As for the last one, it was of a pale, sickly, cadaverous color. "And he who sat on it had the name Death." Death in person, the last of the great scourges over which Jesus Christ, and by extension all bearers of the Gospels, would manage to triumph from on top of a horse that was neither chestnut, black, or pale, but white.

The episode is described by St. John in the New Testament and has been the object of many interpretations since antiquity and throughout the Middle Ages. Whatever the message retained may be, it remains clear for us: in those days, horses must not have been viewed in entirely the same way depending on whether they had a white, a chestnut, or a black coat. Or at least that seemed to be the case in that part of the ancient world dominated by the Christian religion. Did the breeders of those times have those references in mind when they brought a new generation of horses into the world? Did buyers also avoid light-colored horses like the plague, and only used chestnuts when they went off to war? Were black horses never popular? After all, customs other than those gleaned from the Apocalypse could have influenced peoples' preferences, despite the weighty influence of the Church. Didn't the Norman cavalry embroidered on the Bayeux Tapestry following the battle of Hastings show horses of just about all colors? What might the horses of the Middle Ages have looked like, and how can we find out?

If the colors of the four horses didn't change in the Biblical text, we know that the way they were represented underwent several transformations. For example, in the illuminations of the *Beatus of Facundus*, a manuscript copied in 1047 for the king of Spain, Christ's horse appears dappled, not white, and a horse used in war seems quite pale for a chestnut, almost light beige. The colors of our four horses regain their more canonical looks in the *Osma Beatus*, copied around forty years later: the same scene and the same object, but with different colors. Illuminations and painting—graphic art—tell us a story that is rarely devoid of the influence of stylistic fashions and conventions. To take this type of source literally in trying to figure out which animals lived in the past is always to run the risk of perceiving a reality shaped by the choices of artists, narrative constraints, and the ideological context of an age. If on the Bayeux Tapestry the English are on foot and the Normans on horseback, it's because that immediately enables the viewer to identify which soldiers represent which side of the battle. Adding thousands of foot soldiers who also fought on the Norman side on that day of October 14, 1066 would perhaps have more realistically illustrated the historical fact, but would have inevitably made the battle scenes more complicated and more difficult for the artist to execute.

There are other texts we can consult to complete the tableau and obtain a more precise idea, such as the medieval-era chansons de geste, for example. There are also real-world examples that contain royal financial records and official transactions, such as the Pipe Rolls. These are preserved as treasures in the UK National Archives, and for each year between 1294 and 1361 one finds listed the number, ages, and colors of the horses that the crown had acquired, notably at Odiham Castle in the south of England, but in many other places, as well. This type of archive, as well as the cartularies that compile the administrative lists relating to the property of abbeys and cathedrals, are few and far between; they do not cover all periods, and they also present a look at society solely from on high. They do not represent the entire society and do not illuminate all its complexity. Arne Ludwig, professor at the Leibniz Institute for Zoo and Wildlife Research (IZW), came up with a completely different idea for finding out what medieval horses looked like.

DNA and the Coats of Medieval Horses

I met Ludwig for the first time in Basel, where we were both participating in the high mass of biomolecular archeology that takes place every two years somewhere in the world. It was its sixth edition, toward the end of summer 2014. I went to present our most recent discoveries regarding the existence of an unknown horse that was the ancestor neither of modern domestic horses nor of Przewalski's horses—that new lineage of prehistoric horses represented by the Lena horses, as we've seen in chapter 5. But I also went in the hope of meeting Ludwig. I had been following his work closely for several years, because he had been the first to use DNA to determine the color of the coats of ancient horses. He had developed a lab technique for that even before the new generation of sequencers gave us access to entire genomes. His principle rested on a very simple idea: it involved targeting certain very precise places on the genome to copy them into a sufficient number of molecules, in order to read the information contained in their sequencing. Most of the time there was no preserved mummy to evaluate—thus no hair or skin, only skeletons—but the sequence would enable us to deduce the animals' color if there was enough DNA available. The reason is simple: our DNA is the same in each of the cells of our body. Using bones and teeth, DNA made it possible to "see" a horse's skin and its coat. It was incredible.

The genetic instructions that lead to the appearance of a bay-colored coat, a black coat, or a chestnut coat are well known. They rest on the information carried by only two genes, the *ASIP* gene and the *MC1R* gene, which can naturally exist in several versions in horse populations. One of the versions of the *ASIP* gene is shorter than the others by eleven letters (the underlying mutation is called a deletion), whereas different versions of the *MC1R* gene differ from the others by only one letter (the mutations involved are called substitutions). Two mutated versions of the *MC1R* gene are of very particular interest to us here, because the horses that carry the first or second of these versions will always develop chestnut coats. In practice, it suffices to sequence those two letters of the *MC1R* gene to know if the coat of a horse is chestnut, without needing to sequence the billions of other letters present in the genome. That is very convenient.

But there is nothing magic about it. The mutated versions of the *MC1R* gene have the effect of preventing melanocytes, the cells whose specialty it is to manufacture skin pigments, to react to a hormone called melanocortin, which normally influences the production of black pigments. Rendered unaffected by the action of melanocortin because of the versions of the *MC1R* gene they carry, the melanocytes will be condemned never to be able to influence the signal transmitted by the hormone. They will produce no pigments other than the one produced by default: red ("pheomelanin" is its scientific name). The horse will then be red, as the Gospel says (that is, chestnut). The crossing of two chestnut horses will always result in the birth of a horse of the same color as its parents. As each parent carries only the mutated versions of the *MC1R* gene, that offspring will only be able to transmit the mutated version to its offspring. Whether the two parents each transmit the same mutation or a different mutated version doesn't matter; the two copies their foal will carry will always be mutated versions.

The result of the crossing of two non–chestnut horses can, however, be more colorful. If the parents are carriers of a single mutated version of the *MC1R* gene, they will on average have one chance in four of giving birth to a chestnut horse, but the rest of the time they will produce either bay horses (brown manes and black tails) or black horses. This is where the *ASIP* gene (and its version shortened by eleven letters) intervenes. The *ASIP* gene encodes for a protein that normally takes the place of melanocortin and prevents it from acting; in the places where this gene is expressed, which is in all melanocytes except those of the mane, tail, and bottom of the legs, melanocortin cannot exercise its effects, and therefore black pigments are not produced. The horse will develop a coat with red-brown pigments on most of its body, but black pigments (eumelanin) at the mane, tail, and bottom of the legs. A horse will be bay as long as it carries at least one copy of the *ASIP* gene in its normal version, because its product will always be able to counter the effect of the melanocortin in the tail, mane, and lower legs. If instead it carries two copies of the gene in its deleted version, the protein ASIP would never be able to take the place of melanocortin; the effects of the hormone would be exercised unhindered and the melanocytes of the entire body—mane, tail, and lower legs included—would begin to produce black pigments. A horse will be black when it carries two copies of the deleted version of the *ASIP* gene, but bay in the opposite case, except if it carries its two copies of the *MC1R* gene in mutated versions (in which case, it will be chestnut).

What's important to remember is that by determining through sequencing which versions of the two genes the horses of the past carried, our knowledge of the genes' refined biological effects is enough to enable us to deduce the color of their coats (at least for those that were bay, black, or chestnut). This is the foundation of Ludwig's idea: a biological engineering made of types of genetic switches with the scientific names of *ASIP* and *MC1R*, whose combined effects condition the nature and location of pigments produced in the coat.

His idea went much further than that, because its principle could be applied to a much broader spectrum of mutations and enable us to investigate an entire range of other colors. Some involve the *MC1R* gene and can lead to versions of the gene other than those we have seen up to now, notably some whose effects can lead to animals of multiple nuances of white; other mutations can involve genes such as *TRPM1*, and result in dappled horses. Still others act as supplementary genetic switches to modulate colors, spots, and even the presence of stripes on the legs and a black line on the back as in Przewalski's horses (with a dun coat). It is unnecessary to go into the biological mechanics of all this in minute detail, because in the end the principle remains the same: once determined through sequencing, the versions of the genes carried by each individual enable biologists to have an image of what the animals of the past looked like. I invited Ludwig to join our efforts and apply that principle to the collections of horses we had both amassed. To my great delight he accepted, and it was the beginning of a collaboration that continues to this day. While initially we chose to apply his original method and restrict ourselves to a limited number of genes, we later undertook to characterize genomes in their entirety, which nevertheless made it possible to go back to these particular genes to deduce coat color.

What Did the Vikings' Horses Look Like?

The first study we would undertake together involved more than 200 horses. It revealed that coats of diluted shades, such as cream colors, or spotted coats, like pintos, were in clear decline since the fifth century CE and the end of antiquity. Single-color coats, in particular bay, black, and chestnut coats, had been more popular during the Middle Ages, in spite of their gloomy connotations in the Gospels. Over the long term, it seemed that tastes had changed from one period to another and that in the Middle Ages dappled horses were no longer as popular as they had once been; their hour of glory had peaked close to 2,000 years earlier, at the very beginning of the Bronze Age. In this same vein, an anomaly in our data attracted our attention: none of the twenty-one Icelandic horses we had sequenced carried a mutation associated with spotted pinto coats. This was surprising, because those coats, like many others, are relatively frequent there today. But this is also because Icelandic vocabulary, with more than a hundred words to describe horse coat colors, patterns, and shades, isn't lacking. This could mean that pinto horses and the underlying mutations were particularly rare among the first animals the Vikings had brought to the island to establish their herds; in statistical terms, their frequency could not in this case have surpassed more than a few percentage points. They would have become more popular only following changes in tastes that came later in their history. But we might also imagine that the strict ban on importing horses onto the island, which had been in effect since 982, had not been as effective as one might have believed and that the characteristic had entered Iceland as contraband.

The Vikings' horses also had another interesting characteristic: their gait. In addition to putting the *ASIP* and *MC1R* genes into the genetic hopper, with Ludwig we decided to look carefully at another gene with the equally unpoetic name of *DMRT3*. Rather than looking at the complete gene, we were mainly interested in a mutation involving only one letter, located at a precise position on this gene. In 2012 a Swedish group had discovered that this mutation had very serious effects during development and could change the placement of some neurons: specifically those located along the spinal cord that project their nerve endings directly onto motor neurons, whose role is to order certain muscles to contract or to remain relaxed. Such muscles may appear on either the same or opposite side of the body as the neurons controlling them. This detail is important, as we will see. In lab mice completely lacking the *DMRT3* gene, the cabling that connects the spinal cord neurons to the motor neurons and the muscles they control is slightly modified. As slight as it may be, this change has important effects on the animals, in particular on their locomotion. When placed on treadmills, lab mice without the *DMRT3* gene will no longer be able to sustain higher cadences as easily. Breaking down the details of their movements, scientists realized that their gaits were no longer completely the same as those of normal mice (geneticists call them wild types, to distinguish them from mice who did lack the gene). Their limbs spent more time in propulsion and extension, and their gaits were on average longer. Even more surprising, the right legs and the left legs no longer functioned in a perfectly synchronized manner, and the usually well-oiled choreography regulating the extension or flexing phases appeared chaotic. The mice without the *DMRT3* gene seemed to have difficulty coordinating the movements of their front and back legs and their right and left legs at the same time.

There is a mutation that affects the *DMRT3* gene and incidentally shortens the gene's product, and that mutation is particularly frequent in a very specific category of horses: natural amblers, like Icelandic horses or Colombian or Peruvian Paso Finos, which develop a gait in addition to a simple walk, trot, and gallop. Like walking, an amble is a four-beat gait, but the feet on the same side of the horse move forward one after the other, usually in a footfall pattern of right rear, right front, left rear, left front. This is called a lateral movement (the legs of the same side advance together), whereas trotting is a two-beat diagonal movement, because the front left leg and the back right leg advance together followed by the front right and back left legs. The mutation of the *DMRT3* gene, when it is carried in two copies, enables horses to develop the amble naturally. It is a gait that has two advantages. The first is that riders find it particularly smooth, since their center of gravity is not shifted up and down with each step, as it is with the rising trot. The second is that it enables the horse to move fast without galloping. It's important to note that not all horses who carry two copies of the mutation will necessarily be amblers. American Standardbreds, for example, carry only the mutated version of the $DMRT3$ gene. However, their population is equally divided between those that trot and those that amble, and those categories do not compete in the same races. Among standard Swedish horses or French trotters, in whom both versions of the gene exist, the carriers of the mutated version have better racing results. The mutation with its effects on the neurons and their motor projections is permissive for the amble, but it does not necessarily predict ambling; it offers the benefit of a faster trot and a more comfortable riding experience.

Let's go back to ancient horses. Which among them have proven to be carriers of that mutation of the $DMRT3$ gene? It is precisely the horses of the Vikings, and there were twelve of them out of the fifteen that provided us usable genetic results. It seems the Vikings had a true taste for natural amblers (seven out of ten horses carrying the mutation carried it in two copies), as seen in the legendary Norse sagas (*fornaldarsögur*) that tell of the exploits of their heroes, as well as the beginning of the colonization of Iceland. It is easy to imagine that the gait of these horses was especially appreciated in a virgin environment without roads. By sequencing the genes found in bones and teeth, we were beginning to catch a glimpse of what no archeologist had ever hoped to be able to see: the moving animal. The gene is not a dead trace; it tells us about an animal's functional characteristics and its behavior. It provides a way to bring it back to life.

From the Vikings to Genghis Khan

Our analyses indicated that the mutation of the *DMRT3* gene did not originate in Iceland, since it was already found in English horses that lived near York between 850 and 900. The Vikings were able to acquire their favorite horses during one of their many raids on the British Isles before taking them to Iceland—their genomes are very close to the ponies of the Shetland Islands and the horses of the granite moorlands of Dartmoor ponies in southwest England. Another scenario is possible: the mutation might have appeared on the Vikings' native Scandinavian lands, and they brought it to the British Isles and crossed their animals with Indigenous horses. We are not yet certain of this, because the number of Viking horses we have analyzed in Sweden and Norway is still too small for our purposes. One thing is certain: the mutated version of the *DMRT3* gene gradually reached other regions of the world. It has been found in the thirteenth-century site of Tavan Tolgoi in Mongolia, among horses dating from the Great Empire of Genghis Khan. It is possible that the Vikings, during one of their eastern expeditions that reached the Caspian Sea, or others after them, had been able to help this mutation cover a lot of ground.

There is another thing we have learned about Viking from DNA. In Iceland, as in Scandinavia, Vikings didn't sacrifice just any horse to accompany their dead to Valhalla; they almost always chose males. Perhaps this was to assert a patriarchal dominance (in the sagas, calling a man a mare was an insult), or to evoke fertile power through a phallic symbol. Or maybe it was to celebrate the cult of fertility in reverse, by sacrificing the one sex that cannot produce life. We don't know. In any case, the sacrifice of horses was evidently an integral part of Viking culture, ritualized, staged, and dramatized. The horse, generally chosen in the prime of its life, between five and fifteen years old, was saddled and restrained, and then knocked out by a blow to the head before its throat was cut. The spectacle of the outpouring of 20 liters of blood contained in a horse's body might have galvanized a feeling of belonging to the community; the sacrifice didn't just concern the sphere of the dead. Beyond its psychopomp role accompanying the dead on their final voyage it had an eminently political role, and served to unify the group; it therefore equally involved the sphere of the living.

The Medieval Warhorse

A huge steed, covered in iron and ridden by a valiant knight in armor; or a draft horse, perhaps an English Shire or a French Percheron. In any case, a force of nature easily weighing over a ton and measuring 1.80 meters to the shoulder. That is the image used most often to represent the horse of the Middle Ages. Both horses were capable of bearing the weight of a metal-clad rider without flinching, but also raising him up above the melee so he could wield his sword with fearsome efficacy to slay the foot soldiers below him. However, the size of the animal we refer to as a destrier or "warhorse" is still debated among historians. This is because medieval languages were not always as precise as we'd like, and horses were most often described by their use rather than by their breed type—that is, by their standardized physical traits and other biological characteristics. Alongside the very prized destrier of the knights and warriors was the steed or the rouncey, described as lighter and faster. But wouldn't riding a horse of modest size be easier than riding a giant? The rider could then both dismount and easily get back up into the saddle, depending on the circumstances of a battle. The riders embroidered on the Bayeux Tapestry do not appear miniscule compared to their mounts; just the opposite is true. In Arthurian legends, Sir Gawain's horse was named Gringolet (from the French "Gringalet," for "runt"). Don Quixote, the "sad-looking knight," rode Rocinante, a rouncey. So were medieval warhorses, also the most faithful companions of the legendary knights, monsters or runts?

"High medieval destriers may have been relatively large for the time period, but were clearly still much smaller than we might expect for equivalent functions today." This statement, made by Alan Outram to the press at the beginning of 2022, seemed to settle the debate once and for all. Outram is certainly one of the greatest specialists of Botai horses in the world (as we saw in chapter 2), but he also admits a passion for the horses of his native England. A few years ago, Outram and Oliver Creighton, also a professor at the University of Exeter, launched the Warhorse Project, which focused on the archeology of the warhorse in the Middle Ages using the most innovative tools available to determine whether the destrier had represented a revolution in medieval societies. This involved evaluating everything closely or tangentially related to the medieval horse: its stirrups, which were already quite widespread in Europe around the seventh century CE; its bits, saddle, armor, and caparison, the cloth that draped it both as protection and decoration; its skeleton and its pathologies; and its genes, too. In one of their studies, Creighton and Outram assembled the largest archeological collection of horse bones ever collected in England: close to 2,000 pieces from 171 sites spread over the entire territory, from Exeter in the south to Newcastle in the north. This enabled them to measure some dozen skeletal elements to detect variations in the size of the horses, as well as their robustness, from between the fourth century and the middle of the seventeenth century.

In terms of height, the data collected from long bones left little room for doubt; they did not indicate a history of progressive growth, but rather one of a great stabilization until the beginning of the sixteenth century. From late antiquity, spanning the reigns of the Saxons, the Normans, and throughout the Middle Ages, the height to the withers of English horses did not exceed 1.30 or 1.35 meters on average. This would categorize almost all of them as ponies according to the current guidelines of the International Federation for Equestrian Sports, which sets 1.48 meters as the tipping point that turns ponies into horses. These horses in the archeological sense were essentially ponies in terms of their size—an image quite far from that of the familiar medieval behemoth. Only a handful of them managed to exceed that 1.48-meter threshold, by scarcely a few centimeters during the Norman period or around fifteen centimeters in the thirteenth century. More than their size, medieval English horse breeding seemed to have been oriented toward the improvement of horses' other characteristics such as their temperament, endurance, or speed. It wasn't really until after the sixteenth century, once the Middle Ages had passed, that we witness a spectacular increase in size, with close to a quarter of the material analyzed in the study deserving the category of horse, with a height to the withers increasing by around 5 centimeters on average. This was perhaps the fruit of a Tudor policy that aimed to reinvigorate the breeding of warhorses, or it might represent the growing need for work horses during that time for pulling, either in the fields or in towns.

Robustness is a more dynamic situation, in which the horses of the Normans, starting in the second half of the eleventh century and throughout the twelfth century, were distinguished by the morphology of their hindquarters, which appears to have been less stocky than in preceding centuries and in the centuries that followed. Should this be seen as a return to more powerful horses during the twelfth and thirteenth centuries, horses capable of both carrying their rider and pulling new, heavier conveyances? It is difficult to say at this stage, but it is remarkable that it was precisely at that pivotal period that the use of the horse collar began to be widespread in central and northern Europe, thereby giving horses the opportunity to take over from oxen in working the fields. The neck yoke, sized for oxen, was ill-suited for horses, forcing them to pull the weight of their burden from the front, by the chest, at the risk of choking; with the shoulder collar the animal was freed of that handicap and could develop its full power by pushing forward with its hindquarters. It is conceivable that the combination of new equipment and the selection of an animal with adapted hindquarters enabled the farming of more territory than could be done using oxen, and that the foundations for an agricultural revolution were thereby laid. These are in any case the controversial theses that have been proposed. Should we see the hindquarters of the Norman horses as an anomaly revealing the presence of foreign blood, notably that of svelte Arabian horses, which had already begun to infiltrate Europe? These are the types of debates that the analysis of those horses' DNA being undertaken in my lab promises to settle. This is why Outram could not envision a Warhorse Project without our genetic analyses, and why I was delighted to join his ranks.

Horse-Size Genes

Genome studies also have something to say about size. If for humans the genetic factors that influence height remain badly understood, it is because they involve the combined and minuscule effects of hundreds of genes. But the same is not true of horses. For them, only a handful of genes is enough to account for the phenomenal difference in height that separates the miniature Falabella, with its 75-centimeter height to the withers, from the giant Shire and its 1.80 meters. Why? It's quite simple: for our species, there are many, many reasons why one day two people decide to come together and bring children into the world. The genetic patrimony of all types of people is constantly mixing, whether they are brown-haired, blond, tall, short, fat, weak, and so on. With horses it isn't the same, since reproduction is often the exclusive domain of breeders. In particular, since the nineteenth century reproduction between similar horses has become the norm; it is the foundation for creating biological breeds. By being careful to breed large horses with other large horses, small ones with small ones, the fastest with the fastest, and amblers with amblers, breeds characterized by standards and precise specifications have been maintained. The outcome of this is that the genetic pool of horses is considerably limited, and the genetic mutations that cause the appearance of various sought-after characteristics are quickly isolated and favored so that they appear much more frequently.

If breeders are aiming for a tall breed and it happens that certain mutations favoring the animals' growth are present naturally in the founding population, those mutations will be able to infiltrate the budding breed generation after generation. If no mutation favoring the growth of animals is present, breeders will come up empty-handed, but nothing will prevent them from trying again from a new pool of founders. With repeated attempts, breeds with imposing height will emerge looking like French Percherons or English Shires. The same is true of miniature animals like the Falabella, whose history begins in Argentina in 1845 and where these precepts were followed to the letter in the goal of creating a pet horse. One might say that the goal was achieved, since the breed doesn't get any bigger than a large dog—which also contributed to its international success, attracting so much attention that very famous people like the Kennedys or Frank Sinatra acquired them.

At this stage geneticists can take over, and they have breeds of very different sizes available to them. They can measure the size of the animals, sequence their genome, and discover which genetic letters are overrepresented in tall or short animals. They can then determine the key genes that govern the size of horses and the average degree to a mutation will affect the size of an individual, making it bigger or smaller. It is exactly this principle that was applied in a study that appeared in 2012, which claimed that only four genes (or close to that) were enough to explain most of the variations in size observed in horses, or about 83 percent of the overall range. In other words, the portion of the innate was predominant in size differences, since other factors, such as nutrition and the environment, played only a marginal role and were able to explain only 17 percent of the measured differences that remained.

The discovery was sensational, because on the scale of individuals (and no longer breeds), the information carried in only four letters out of the 3 billion or so contained in the genome of a horse was enough to increase the height to the withers by several centimeters. Another study published at the same time found two of these genes by starting not from variations among breeds, but from variations within a single traditional Swiss breed, the Franches-Montagnes or Freiberger. Here, the innate portion quite obviously played a less significant role insofar as animals that were all very similar genetically were being compared. Since then, many other studies have been guided by the same principle and ended up with approximately the same genes, at the forefront of which are the *LCORL* and the *ZFAT* genes. Some studies have even identified mutations that affect both the height and appearance of the animals, such as the size of their hocks, chests, and their overall build.

We have veered quite far from the horse of the Middle Ages, but we have also learned something essential: we can use DNA to measure the size of ancient horses, and this is possible because whether we find a piece of a tooth, a bone, a rib, or a bone from the skull the cells of an individual, all share the same DNA. We can even use it to measure how the breeders of the past did or did not focus on selecting increasingly larger horses: the faster the mutation associated with increasing size spread within a population, the greater the breeders' intervention, and the more obvious their selective activity. This is why the genetic approach is beneficial when used alongside that of Outram and Creighton, since their approach requires a caliper and a very particular type of bone like femurs, which we call long bones and are not always preserved archeologically.

Before unleashing our genetic arsenal, we still had to verify that the mutations implicating the *LCORL* and *ZFAT* genes were as universal as they appeared. Though most studies seemed to point in that direction, they had mainly focused on breeds of European origin. It could be that the history of breeding in Asia had led to the selection of other mutations governing significant differences in size there. We were therefore going to borrow the same principle, but this time apply it only to breeds of ponies and horses of Asian origin, mostly Chinese. It was the right move; we found that a difference of only one letter in a region affecting the expression of the *TBX3* gene could account for a good portion of the differences in size observed in Asia. In lab mice, we could even modify the region of the genome equivalent to the one carrying the identified mutation and demonstrate its spectacular impact on how fast the increase in the size of limbs occurred during development. To follow the changes in size that breeders of the past might have selected in Europe and Asia, we had to look at the letters carried by the genomes of horses for three genes: *LCORL*, *ZFAT*, and *TBX3*.

This work indicated that the mutation associated with *TBX3* that increased the height of Asian horses had begun to become more frequent much earlier than in the Middle Ages: at the end of a period called in China the Warring States period, and until the first Imperial Dynasty from which the country was named—the Qin Dynasty. This was around 2,200 years ago. Since then, the genetic history has been one of a mutation that continued to spread gradually within the population, as if the demand for taller horses had never decreased. In Europe, mutations associated with taller horses also began to gain ground around the end of the Iron Age and the beginning of the Roman Empire, without reaching the same proportions as in Asia. Even if the data still need to be consolidated, the frequency of the mutation involving the *LCORL* gene seems to have had only very modest gains during the Middle Ages, which would seem to corroborate on a continental scale what Outram and Creighton observed in the bones of horses of England. It doesn't appear that creating tall behemoths constituted the main preoccupation of breeders of the Middle Ages, in spite of the reputation of medieval destriers.

City Horses, Country Horses

Before leaving the Middle Ages, let's look one more time at what we have recently discovered. We are now back in France, where we have already analyzed the remains of close to a thousand horses from the period spanning the end of the Iron Age up to the nineteenth century. From all periods, in particular during the Middle Ages, archeological sites both in urban settings and in rural castles all contained a larger proportion of male horses than of female horses. We know this because DNA doesn't just allow us to determine colors or height but also sex, as we have seen. It appears that mares lived in the countryside, around villages and abbeys, whereas males went to cities very early in their lives. This is likely because the countryside offered an environment more conducive to breeding than stressful, noisy cities, and because the presence of females in heat would have no doubt distracted stallions from their work, unless they had been castrated.

However we interpret the data, this discovery shows that horses represented precious resources for which neither their breeding nor distribution were left to chance, but instead reflected true rational management. This study invites us to dig deeper. Were the males that were preferentially used in cities all of the same type, multitaskers capable of assuming all duties? Or did different types coexist, each specializing in the execution of specific tasks? If that were the case, what were their geographic origins? Was breeding regionalized, with each region producing its characteristic type or types? And if this was the case, on what scale: France, Europe, beyond? Of which modern breeds are medieval horses the ancestors? Were they for example genetically closer to "cold blooded" horses, known for their quiet temperament, or to the more spirited "hot blooded" breeds? Beyond temperament, what were the physical and genetic characteristics that were most prized? Did those preferences, if there were any, stay the same throughout the Middle Ages, or did they change? When? Why?

There are enough questions to make one dizzy, and to provide enough fodder for at least an additional decade of interdisciplinary research. That research will be based on paleogenetic approaches; DNA will enable an investigation of the demographic dynamics and diversity of color, height, and type in a particular place and time. It will also be enriched and reinforced by many other approaches, such as isotopic research; in a city, animal carcasses were often removed and piled up in rendering centers, which today constitute the main source of archeological material available. The animals that finished their days in such places were not necessarily born near them and had perhaps lived a thousand lives before ending up there: the life of a warhorse, a farm horse, a pack horse, or even that of a steed. The analysis of isotopes—atoms of the same element with different numbers of neutrons (see chapter 2)—contained in tooth enamel can inform us about what an individual ate and about the environment it encountered during its life. Isotopes in baby teeth or in various bones can also provide information about the place where an animal began its life and the one where it ended it. It would be interesting to compare such information with that derived from anatomy, which informs us about how an animal was primarily used and about its health, and the information derived from the historical and economic sciences, in order to assess all the decisions that were in play in the reproduction and management of horses.

Alan Outram and Oliver Creighton were quite inspired in their attempt to make the Warhorse Project as integrative and collaborative as possible. It is now up to us to rise to the challenge and be inspired by them to complete the portrait that English sources provide with what other sources can tell us, such as French sources. We will then obtain completely new insight into the way in which these two powers, rivals in ancient times, were able to draw out the best part of the horse to enhance their armies and build their economic wealth at critical moments in their history.

9

The Horse of Extreme Lands

The Far East and Its Legendary Tea Road

The Ancient Tea Road is not as well-known as the Silk Road, but its raging rivers in rugged canyons and steps dug right out of the rock, clearing a difficult path on the sides of mountains among the highest in the world, are no less legendary. This road was once taken by caravans of merchants who transported tea, sugar, and salt over close to 2,500 kilometers through China and Tibet. The journey might have begun in the Sichuan province to transport the green tea that is grown there, but it might also have originated in neighboring Yunnan for an even more prestigious cargo: pu'erh tea, a dark tea which, like good wine, is reputed to improve with age. Whatever their departure point, the caravans always followed the same route, and along the way encountered more than twenty different ethnic groups before reaching Lhasa, the capital of Tibet located at over 3,600 meters, four to six months later (if all went well). Before arriving it was necessary to cross some fifty rivers; navigate fifteen or more rope bridges, one more dizzying than the next; climb no fewer than seventy-eight mountains more than 3,000 meters high; and confront many situations when the weather could change in a snap and pit travelers against the harshest storms imaginable.

This road is known by another name: the Ancient Tea Horse Road, to acknowledge the animal without which such an odyssey would have been practically impossible. It was most often mules who, along with yaks, played the role of beasts of burden, guiding the caravans and bringing to port their cargo of precious goods. But what the moniker reminds us that horses, as much as tea, served as currency of exchange. The kingdom of Tibet paid for the shipments of tea not in the coin of the realm, but in horses: one horse for a set quantity of tea.

Official history says that the Tang dynasty in China introduced the Tibetan aristocracy to the benefits of tea beginning in the seventh century CE, but that it was only three centuries later, with the Song dynasty, that a large-scale trade was really established. In exchange for their leaves with miraculous properties, which they delivered in humongous quantities, the Song demanded a no-less-phenomenal number of horses. A study has estimated that in barely more than a century and a half, between 960 and 1127 CE, half of the total tea production of Sichuan was exchanged in Tibet for more than 20,000 horses—representing around 15,000 tons of tea! And that was nothing compared to the volumes exchanged under the Ming (1369–1644) or the Qing (1644–1911) dynasties, periods when the Ancient Tea Horse Road was the most prosperous. In 1661, caravans departing from Yunnan could transport 1,500 tons of tea in a single year. Today, highways and modern modes of transportation have taken over from the road, but it still provided important services during World War II, when Japan occupied China and controlled its modern paths of communication.

The Origin and Originality of Tibetan Horses

By obtaining so many horses, the Chinese authorities were seeking above all to consolidate their military power. Those animals provided them precious support in counteracting the emergence of hostile powers, notably the nomadic peoples of the steppe to the north of the country, who had an endless supply of horses. Were Tibetan horses simple horses like any others, necessary cannon fodder and nothing else? Or, since they came from a region located at the top of the world, did they have something special that their cousins of the plains didn't have? The question deserves to be asked because we know that very high-altitude environments are not really conducive to life: it is very cold, the air lacks oxygen, and the ultraviolet rays from the sun are stronger due to the altitude, where the atmospheric layer that normally protects us is thinner. Despite all that, there are many domestic animals that live alongside horses on the Tibetan Plateau. Yaks, for example, have lived there for a very long time, but also pigs, sheep, goats, chickens, and dogs. The Tibetan Plateau is the widest and highest on the planet, at an average of 4,500 meters in altitude.

I met Xuexue Liu in 2018 while she was working on her PhD thesis at the Institute of Animal Science in Beijing. We were attending the biennial conference organized by the Havemeyer Foundation, which brings together scientists interested in the horse genome. That year we were in Pavia, Italy. I came to present our latest findings in our research on the cradle of horse domestication. Liu was presenting the results of a preliminary study that she had carried out with Lin Jiang, a professor at the Institute for Animal Sciences in Beijing, whom I'd had the honor of meeting a year earlier during a conference in Dublin. Their study focused on the horses of Tibet. It was still incomplete, and Jiang and Liu needed someone to help them finalize the work. We ended up collaborating on a study involving the analysis of 138 horse genomes that was published a year later. Soon after, Liu joined my laboratory to keep working on horses and their evolutionary history.

We discovered that the horses of the Tibetan Plateau—those from Tibet itself and those of the Qinghai province—are not exactly genetically the same as the horses native to provinces in the southwest of the country, including Sichuan and Yunnan. They define two of the three large genetic groups populating China today, with the third being related to horses from the provinces of the north, which, from the entrance to Kazakhstan to the Pacific, merged with horses from Mongolia. This situation seems to have begun very early in the history of the country, quite likely with the arrival of the first domesticated horses in the region in the middle of the Bronze Age, around 3,500 years ago. The genetic data agree with archeological discoveries indicating the presence of horses on the high Tibetan Plateau during the second millennium BCE.

However, the very fact that we could recognize these three large genetic groups in the populations that live today indicated something we weren't necessarily expecting—that despite the considerable number of horses that traveled over the Ancient Tea Horse Road in the return direction for more than 1,000 years, the local herds of Sichuan and Yunnan did not taper off. If that had been the case, the various groups with multimillennial origins would have blended together and would no longer be genetically distinguishable today. The Tea Horse Road nonetheless left some visible footprints in the genetic data. The Tibetan horses and those of Sichuan and Yunnan exchanged more DNA material than they did with Mongolian horses, despite the countless nomadic hordes that spread out over the land, of which those of Genghis Khan and his descendants are perhaps the best known.

What our study also revealed is that of the three large groups identified, the horses of the Tibetan Plateau had been the most prolific, and by far. It is the genetic group that shows the largest effective size, and it also seems that life at high altitude did not really pose any problems for them. Evidently, they must have adapted to that environment. We needed to measure physiological blood variables to try to learn more. The results showed that their red blood cells are individually larger, but collectively occupy a smaller volume in their blood than in that of their low-altitude cousins. That might be surprising, since red blood cells are the ones that convey oxygen in the blood, thanks to the hemoglobin they carry. So how do Tibetan horses not suffer from oxygen deprivation, when they live in an environment where oxygen is already sparse? According to our measurements, this is because the concentration of hemoglobin carried by each of their red blood cells is much greater than in breeds native to low-altitude plains. Tibetan horses benefit from a physiological trick that enables them to have more oxygen in an environment that cruelly lacks it, without running the risk of thickening their blood with more cells and forming clots that could obstruct the blood vessels and rapidly lead to complications such as phlebitis, with the risk of producing often fatal pulmonary embolisms.

The *EPAS-1* Gene and Its Role in Adapting to Life at High Altitude

An analysis of genomes enabled us to understand the origin of this adaptation. At the top of the list of the areas of the genome where Tibetan horses and low-altitude breeds present the most marked genetic differences is the *EPAS-1* gene, well known to evolutionary geneticists. This gene was described fifteen years ago, and it has the distinction of existing in a special version in the Tibetan people, providing their exceptional ability to live in an oxygen-poor environment. The same is true of the dogs in the region, as well as the sheep and pigs. By finding *EPAS-1* on the road of Tibetan horses, it seemed that we were on the verge of uncovering a remarkable phenomenon of evolutive convergence in which independent species during evolution followed the same adaptive trajectory when they were exposed to the same environmental conditions. Furthermore, the version of the gene that is present in close to four out of five horses living above an altitude of 4,000 meters became gradually rarer when we went to less elevated altitudes; it was found in only two out of three horses between 3,000 and 3,500 meters, and around one out of four at the foot of the plateau, in Yunnan. It seemed as if the animals that carried the mutated version of the gene had been favored in the course of evolution as the oxygen deprivation they had to confront became more pronounced. In Darwinian terms, the carriers of the mutation had a greater chance of surviving in that environment and reaching the age of reproduction, when they could transmit their mutation to following generations. That mutation could gradually increase in frequency in the populations concerned in a way that was even more marked, since the environment at high altitude gave it an advantage.

But what was the effect of that mutation, and how did it affect the horses' physiology? At this stage, there are still shadowy areas we are working to clarify. What we know is that when oxygen is lacking, the protein that the *EPAS-1* gene produces returns to cells' nuclei, where it pairs with another protein called ARNT. This couple, which is called heterodimer, forms an activation factor of many targeted genes recognizable because they carry particular genetic sequences at the level of their promoters. The result of this series of microscopic events is that other genes are activated to produce the proteins they encode: such as EPO (erythropoietin), which stimulates the production of red blood cells, and others, such as LDHA or VEGFA. What interests us here are their effects. The product of the *LDHA* gene acts on a major intersection in an organism's metabolic balance and directs reactions in a mode that enables cells to recover energy without needing oxygen; our measurements confirmed this, as the blood of Tibetan horses contained high levels of LDH, the product of the gene in question. The product of the *VEGFA* gene stimulates vascularization and the production of all blood cells, which include the red blood cells. *EPAS-1* acts a bit like a sentinel that detects a lack of oxygen on a cellular scale and unleashes a set of cascading reactions that enable the organism to confront a physiological situation that, if nothing were done, could put it in danger.

Our lab experiments have shown that the version of the *EPAS-1* gene carried by Tibetan horses is unique in the animal kingdom, and that the protein it produces has the property to pair with the ARNT factor in a stronger and more stable way. At the end of the chain, the genes recognized by this pair are stimulated more and begin to produce the proteins they encode in greater quantities, such as LDHA and VEGFA. Through a domino effect, this leads to an amplified response by the organism that will not only mobilize the metabolic pathways that require less oxygen, but also change its blood formulation to compensate for the lack of oxygen. It then produces larger red blood cells richer in hemoglobin, which improves its ability to maintain a basic oxygen supply for all the cells in its body. As we can see, not every mutation is a bad thing.

Strikingly, other mutations in the *EPAS-1* gene have been described in other animal species living on the high Tibetan

Plateau, and they produce similar, convergent effects. The convergence here is functional: the effects of the mutations are the same, not the mutations themselves. Even if it is the same gene involved, we don't find it in the same form. We also know that the version of the gene carried by Sherpas comes to them straight from another branch of the tree of human evolution: that of the Denisovans, whose very existence we were unaware of until the DNA preserved in a child's pinky bone revealed to modern science that at least 80,000 years ago another group of humans lived in the mountains of the Altai. How the Denisovan mutation found itself in the ancestors of the Sherpas, and why the latter went to the Tibetan Plateau 9,000 years ago, thereby exposing the mutation to a selection that hasn't weakened since then, is another story that does not involve horses. But it is clear that as with horses, it is the passage from one environment where the mutation would confer no advantage—the low-altitude plains—to an environment where it proved decisive—high altitude—that imposed the foundations of their biological adaptation.

Horses from the Extreme Cold of Siberia

The horses of Yakutia are just like the horses of the Tibetan Plateau—forces of nature. They do not, like Tibetan horses, confront altitudes that are among the highest on our planet; their feat is to live in latitudes that count among the highest in Siberia. They live in the coldest land in the Northern Hemisphere. In Yakutsk, the capital, thermometers register average temperatures of around -40°C in the months of December and January, and they quite regularly fall below -60°C. In 1891, a record year, they fell below -64°C. This is why that city deserves its reputation as the coldest city in the world. There is worse: Verkhoyansk, a small town with a thousand or so inhabitants located around 500 kilometers north of Yakutsk, shares with Oymyakon, a village on the banks of the Indigirka River, the record for the lowest temperature in an inhabited region ever recorded on earth: -67.8ºC. In these conditions, it is not surprising that close to 40 percent of the ground of this territory, larger than Argentina and barely smaller than India, never thaws—it is permafrost—nor that mammoth hunters today make it their preferred terrain to find the carcasses and coveted tusks of their favorite animal.

Nevertheless, it can happen that the ground warms enough in the summer and is able to nibble at the permafrost a bit—after the trees of a forest are cut down, for example, and their shadows no longer cover the ground during the short summer months. Bacteria then begin to awaken and feed on particularly abundant organic matter. This activity hastens the thawing of the ground and its degradation a bit more. In the long term the process can have spectacular effects, as in Batagay, more than 600 kilometers north of Yakutsk, where many years ago a rather impressive crater was found. That crater, known as the "gateway to hell," did not appear as a result of a celestial catastrophe such as a meteorite crashing to the earth; instead it is the result of local climate effects, initiated by our own activities. The consequence of clearing the taiga forests in the 1960s was enough to begin the formation of a depression, today more than a hundred meters deep and around 1 kilometer long, and growing larger every year.

The locals also refer to the crater as the "gateway to the Underworld," because with erosion its sides collapse and reveal the carcasses of animals from the past. One of those revenants appeared in the headlines in May 2018: a foal barely two months old that had remained frozen for more than 42,000 years. Scientific teams from Siberia and Korea were able to collect a few milliliters of its blood and started to float the idea of cloning the animal one day. However, the images provided to the press were quite impressive by themselves; in the close-ups of the animal's head and in particular of its nose, the detail of the hairs appeared so alive that one might have thought it was still breathing.

The Batagay Horse

I didn't have the opportunity to work on that 42,000-year-old carcass, but I had access to another, which also came from the bowels of Batagay. Its DNA was so perfectly preserved that we really didn't have any trouble producing a high-quality genome sequence. The animal carried an X and a Y chromosome, so it was a male. Radiocarbon dating told us that it had lived almost at the same time as the Botai horses, 5,200 years ago; it might have even crossed paths with them. However, on the genetic level it didn't share much with them, nor with the lineage of modern domestic horses, the DOM2, who didn't begin their unstoppable expansion throughout the world until a millennium later. Instead, the animal's genome told us that it descended in a direct line from *Equus lenensis*, the famous Lena horse that has disappeared today (see chapter 5). It represents the last of its kind whose genome we have sequenced—which doesn't, however, mean that it was the last of the survivors. Adapting over millennia to the glacial cold of these latitudes, the Lena horse could very well have been able to continue to roam the Siberian permafrost for millennia after our specimen from Batagay closed his eyes for the last time. Local legends have it that the horse we find today in Yakutia is the descendant of a population of wild horses that were domesticated on-site a very long time ago.

To settle the issue, we had to sequence the genome of horses living there today. After all, it was easier to do that than to sequence the genome of a multimillennial specimen, no matter how well-preserved it was. Fortunately, my colleague Andrei Tikhonov, from the Russian Academy of Sciences, was able to send me the hair of some dozen animals before winter took over and seriously complicated the logistics of any scientific expedition in that region. Yakutian horses are not bred in captivity; they are left in semi-freedom in the taiga and the tundra, where they wander before being gathered together once a year. Carole Ferret, a researcher at the Social Anthropology Laboratory of the Collège de France, estimated that a Yakutian horse breeder on average spent five to nine days a year caring for his horses. That seems like nothing compared to the care we lavish on domestic horses in our latitudes. The Yakutian horse is small and stocky, with a long, thick coat, and seems to be made-to-order to confront the cold of the region. It also has the ability to accumulate fat in a record amount of time, in the short period of two months when plants can grow. And it has another exceptional asset: it is able to slow down its metabolism in the winter during the extreme cold, without having to hibernate.

Since it took several months before the package Tikhonov sent reached me, I had in the meantime been able to obtain archeological specimens dating from the nineteenth century. They came from the digs that Éric Crubézy of the Université Paul-Sabatier in Toulouse has been carrying out in that region almost every summer for around fifteen years, and consisted of animal remains that had been placed as sacrificial offerings in human graves.

The True Origin of the Yakutian Horse

The analysis of the genomes was conclusive and put an end to the legends; none of the specimens analyzed had much in common with the specimen from Batagay. They all appeared to be full-fledged members of the lineage of the modern domestic horse, the DOM2, whose roots go back to the western steppe of Russia, 4,200 years ago. Instead, the genetic information agreed with the history books, which attributed relatively recent origins to the Yakuts and their horses. Most sources agree that a horse-riding people who occupied latitudes more to the south of Lake Baikal would have initiated a migration north starting in the thirteenth century. Those migrants, who were fleeing the unstoppable surge of Genghis Khan's hordes at the time, would have settled not in virgin territory, but in a place that had been populated before them. They would have laid the ethnic foundations of the modern Yakut people and the cultural foundations of a civilization that Ferret calls the "civilization of the horse." In Yakutia, the horse is not just that national hero flying on the flag of the Republic of Sakha. It is not only that indispensable vehicle in a vast territory that seems to have no apparent geographical boundary. In Yakutia the horse is much more: they eat its meat and drink its milk; they recycle its hide to make clothing and its tendons to make ropes; it is celebrated as the subject of tales and songs. The animal is an integral part of the local way of life.

But if the Yakutian horse didn't descend from the horse of Batagay, was it nevertheless possible that it carried some of its genes? The idea wasn't so ludicrous; close to 2 percent of the genome of people who live in Eurasia today descend from Neanderthals with whom their ancestors mixed. Some of the Neanderthal versions of our genes were even beneficial in crossbreeding and have been able, through natural selection, to be passed down to us—population geneticists call this adaptative introgression. For example, some genes conferred upon Neanderthals the immunity against viral and bacterial pathogens that existed in Europe, and against which our *sapiens* ancestors newly arrived from Africa had no adaptive defenses yet. Was it then possible that a similar phenomenon had occurred in horses in Yakutia? If the Lena horse had not yet died out in the thirteenth century, could it have mixed with the modern domesticated horses that the first Yakutian riders brought with them? Was it possible that these animals inherited their resistance to the extreme climate of the region from the horses they would encounter, which had lived tens of thousands of years before them on the same territory?

Our analyses refuted that scenario. The genetic text carried by contemporary Yakutian horses, like those of the nineteenth century, is not enriched with aspects that would be characteristic of the text carried by the Lena horses; we don't really find more of it in them than in any other modern domesticated horse elsewhere in the world, today or in the past. Contemporary Yakutian horses owe their biological adaptations to the genetics of their ancestors from the thirteenth century and to nothing else. We might then think that the Lena horse had perhaps already disappeared, since the blending seems never to have occurred.

Even if our data confirmed that only a small number of modern domesticated horses had reached the latitudes of Yakutia to establish the current population there, it is still true that they collectively carried a pool of genetic mutations on which natural selection had carried out its work, fashioning the biology of the animal according to the demands of its environment. Some of those mutations considerably improved the chances of survival of those carrying them, giving them more opportunities to transmit them from generation to generation, so that today they are quite common—they are sometimes the only versions present in those latitudes. They affect the genes in the part coding for proteins as much as the promoters, and have involved both qualitative changes in the products of the genes and quantitative changes controlling their level of production. Even if subtle biological mechanisms still remain to be decrypted, the genetic changes thanks to which the Yakutian horse is so well adapted to its environment involve genes with very diverse biological effects, going from the development of hair and its density to the stocking of fat, and including the metabolism of sugars and the regulation of the biological clock that indicates to our cells the length of day and night.

It seems then that evolution did not provide the Yakutian horse with a supergene that would have endowed it with a single and unique superpower, but that evolution proceeded in the species through the coordinated adjustment of a set of quite varied functions. The irony of history is that in this genetic diversity we found some genes that also contributed to fashioning the biology of other species coping with the same Siberian environment, such as the woolly mammoth and even our own species. Thousands of kilometers from the Tibetan Plateau, we once again came nose to nose with this now familiar phenomenon: evolutive convergence.

10

The Horse of the Americas

The Indigenous Point of View

"My name is Dr. Yvette Running Horse Collin, and I am the executive director of Sacred Healing Circle, a nonprofit organization that focuses on supporting American Indian programs and communities." This email appeared on my computer screen one August evening in 2018, a few months after we had revealed the true identity of Przewalski's horses and their direct link to those of Botai. "I am also the researcher behind the attached May 2017 PhD dissertation titled *The Relationship Between the Indigenous Peoples of the Americas and the Horse: Deconstructing a Eurocentric Myth*." I wasn't familiar with that work, as it is grounded in a methodology I know little about, inspired by investigation practices used in ethnography and social anthropology. Fortunately, the rest of the message offered a detailed debriefing. "Although Western academia claims that the horse became extinct throughout the Americas during the last Ice Age period (between roughly 13,000–11,000 years ago), many American Indian Nations contend that they always had the horse, and that the horse brought over by the Spanish conquistadors in the late 1400s and early 1500s bred with the Ancient Horse of the Americas. This merging

of two species serves as the foundation for the wild horse population throughout the Americas today." She summed up by stating
that Sacred Healing Circle and their partners wanted to collaborate with my team on the continued research and preservation of
what was left of these ancient horses.

To imagine the famous mustang—the symbol par excellence
of the endless plains of the American West—as an ancestral horse
of mixed blood was counter to all prevailing theories in Western
academia, as Collin was well aware. Paleontologists long ago established that the Americas are the original cradle of the family
of horses (the equids) that also includes zebras, donkeys, and
khulans. There is no doubt about this—so much so that explaining the evolutive history that spans over 60 million years, starting
with an animal the size of a dog and ending with the large herbivores we admire today, is an obligatory passage in books dealing
with the horse. This history generally provides material for an
opening chapter to a book, with a succession of anatomical and
dental images, with titles most often directly inspired by the
translation of the scientific name of the very first fossil forms observed: *Eohippus* (or *Hyracotherium* for purists), the horse of the
dawn. We'll offer an abbreviated version of that history here.

The Americas: Cradle of the Horse Family

Paleontological clues indicate that *Hyracotherium* left its original
cradle in the Americas more than 50 million years ago; it then
went to the Old World and spread other lineages that were subsequently transformed. This history repeated itself several times
from other American descendants, and as a result of the locations
and environmental conditions they encountered, led to a flourishing of new biological adaptations. The most spectacular transfor-
mation without any doubt involves the anatomy of the legs, whose

extremities first rested on several digits in front and in back; some tens of millions of years ago this reduced to only one digit, thereby endowing our equids with an anatomy that is more familiar to us: limbs ending on hooves. We must imagine for the horse family a particularly lush evolutive tree, punctuated with periods of great radiations both in terms of geographical expansion and diversity of species, while also experiencing episodes of extinction when some of those forms disappeared forever. In that evolutive ballet, *Equus* represents the last large group of equids to have emerged, around 4 million years ago. It is also the only one that still exists on our planet today. Scarcely 10,000 years ago there still existed other genera of equids, with radically different morpho-anatomies. They ultimately died out—like the *Hippidion* of South America, for example, which had a prominent and rounded nasal profile resembling that of the Mongolian saiga antelope, whose stocky limbs contrasted with the particularly graceful and long legs of the *Haringtonhippus*, a close but distinct genus from North America.

Equus had also been established in North America. Very early in its history it led to two main lineages: the one that ultimately produced horses—the caballine lineage—and the one that resulted in zebras, donkeys, and khulans—the stenonines. This great divergence started around 4 million years ago and finished between 2 million and 3 million years ago, if we are to believe the number of genetic differences we find today in the genomes of these species. It quite likely occurred when an ancestral population succeeded in reaching the coasts of Siberia as a result of a lowered sea level. While continuing to travel across Eurasia and then to Africa, the lineage gave rise on the way to the diversity of the stenonines that we know today.

The caballines were not to be outdone; they also crossed the Bering Strait during another lowering of sea levels, somewhere around 800,000 and 1 million years ago, according to the same genetic estimates. They then led a double life, divided between the Ancient and the New Worlds. In the Eurasian territory their evolution led to the emergence of a large number of horse lineages, one of which was domesticated in Botai but ended up being replaced by the one that was domesticated in the lower Volga-Don valley 4,200 years ago, as we have seen in previous chapters. In the Americas, the evolution of caballines also surely led to the emergence of many lineages and turned out to be a naming nightmare exercise for paleontologists who endeavored to describe the diversity of equid types revealed by their discoveries in the field. However, we know little about them on the genetic level, with the exception of those found on the latitudes of the Great American and Canadian North: from some of those, dating around 700,000 and 28,000 years ago, we have been able to characterize the complete sequence of the genome, as well as the sequences of more than a hundred mitochondrial genomes corresponding to the part of DNA that is transmitted in mammals only through the mother. This work is the culmination of a long-standing collaboration with Beth Shapiro, a professor at the University of California in Santa Cruz, and the author of a mammoth-cloning manual.

Exchanges from One Continent to Another

In addition to clarifying the timeline of the evolutionary history of caballines in the Americas, our collaborative work revealed that the lineages of the two continents did not remain unaware of each other, but came into contact several times after they split. If the first journey had taken place from the east (the Americas) to the west (Eurasia) and had initiated the first great divergence, other journeys were repeated after that. The typical mitochondrial DNA of those animals found in Eurasia reached the Americas somewhere between 130,000 and 60,000 years ago, indicating that mares had completed their journey east and transmitted a genetic type originating in Eurasia into the heart of the Americas. In contrast, some mitochondrial genetic types to this day have never been found outside the Americas and thus reveal the existence of Indigenous lineages on this continent, such as animals associated with distinct paleontological forms like *Equus scotti* or *Equus lambei*. All of this indicated incontestably that the horse, not only as a family (equids) but also as a species (*Equus ferus* or *Equus caballus*), was born in the Americas, had reached the rest of the world, and had even been able to return to the Americas following major global climate changes—such as those great glaciations that froze water in the polar caps to lower sea levels and make the Beringian land mass emerge, thereby bridging America and Eurasia together. The horse has been in the Americas since time immemorial, well before our own species appeared 200,000 to 300,000 years ago in Africa and ultimately in the Americas around 20,000 years ago (or even earlier, depending on the sources one consults). Within the deep past, the horse and some of the ancestors of Yvette Running Horse Collin—whether those of the Lakota tribe like herself as well as the famous war chief Tȟatȟáŋka Íyotake (Sitting Bull), or of those from other Indigenous tribes—shared the same land, and they had ample opportunities to become familiar with each other. So far Collin's theory held up, and didn't contradict any of the data of Western science.

Horses Return to the Americas: The Last Hints of Colonial Eurocentrism?

It was in what followed that the two stories clashed, because the last horse fossils ever found on the American continent date from around 13,000 years ago at the latest. If horses had continued to populate the Americas after that date, we could expect for them to have left other traces up until the continent was rediscovered by Christopher Columbus and colonization began to rage. This lack of paleontological evidence is the main argument advanced by Western academia to support the hypothesis of a horse extinction on the American continent shortly after the last great glaciation, with horses only returning at the beginning of the sixteenth century in the holds of European ships. According to that argument, the horses of the Americas, whether they live in freedom like mustangs or are kept in pens, could not be the mixed breeds that Collin described in her thesis. However, she relied on sources other than fossils: an unprecedented work of memories gathering the long-ignored and even silenced accounts of Indigenous Americans recounting their connections to the horse. The transfer of traditional knowledge among these peoples was oral, and for the most part Western observers didn't begin to bring back their own accounts about horses until the end of the seventeenth century. This is why Collin used the term "Eurocentric" in the title of her thesis. She drew on traditional knowledge to reject the authority of the paleontological argument. In her opinion, the absence of fossils, as crushing as it might have seemed at first glance, could have very well revealed the existence of a confirmation bias among paleontologists; the latter might be so convinced of Western dogma that they would immediately classify every horse fossil found in recent geological strata as colonial. That same absence could also reveal the existence of populations in decline, which could hardly be expected to have left thousands of fossil traces. But it's important to note that an absence of fossils does not necessarily mean extinction.

Sedimentary Traces of DNA More Recent Than the Widely Accepted Age of Extinction

The argument might appear hopeless if traces other than fossils hadn't begun to indicate the possible persistence of horses in America several millennia after the official date of their extinction: 10,500 years ago, and perhaps even 3,000 years later. These traces are not visible to the naked eye but are molecular—again, DNA. In the absence of fossils to study, scientists decided to direct the power of their instruments at sediments in the hope of finding the ghosts of horses—looking for traces of a past that had literally turned back to dust, in other words. Yet when professor Eske Willerslev of the University of Copenhagen had this brilliant intuition in 2003, before sequencers had become the powerful monsters they are today, he found genetic traces of extinct animals, including woolly mammoth in sediments in Siberia. He also found traces of moas in cave sediments from New Zealand DNA—those birds incapable of flying that still populated the archipelago at the time of its discovery by Abel Tasman in the 1640s, but which are now extinct.

The technique had thus proven itself and Willerslev, who was my mentor at the University of Copenhagen for the ten years or so that I worked there, was happy to apply it to sediments of North American permafrost he had extracted north of Fairbanks, Alaska. Among the sequences discovered in a stratigraphic layer dated between 10,500 and 7,600 years ago appeared, alongside those of mammoths, bison, and moose, that of an *Equus*, maybe even a horse. This genetic trace seemed restricted to the stratigraphic layer of interest, and did not show up in the sediments located above or below. It was difficult to believe that it could have derived from a modern horse through surface excrements contaminating the precious stratigraphic layer, and it seemed even more difficult to claim that it derived from much older animals and had somehow defied gravity to percolate up through the sediments. We could therefore reasonably believe that the horse's death knell had perhaps not sounded according to the exact terms of the dominant paleontological model. In the very ranks of the cutting-edge Western science she challenged, Collin found unexpected support—at least if it were confirmed by other genetic studies carried out on the sediments.

This is what was done later in several places on the North American permafrost, with the most recent equine sequences obtained to date estimated at around 7,600 and 8,000 years ago. These traces are admittedly much less frequent than in the older sediments, and sometimes even absent for millennia before reappearing sporadically, which indicates a population in clear decline and surviving in fragmented habitats rather than thriving around the continent. As a comparison, moose and reindeer seemed to have been much more common at the time, according to these same genetic proxies. The entirety of this work suggests that horses could have survived in America until the middle of the period that geologists refer to as the Holocene. That piece of data is enough by itself to revolutionize our understanding of their extinction, if there was one, and to disqualify the model called a *blitzkrieg*, which stipulates that the demographic surge of local human groups that gained ground at the end of the last glaciation would have led to an almost instantaneous extinction of many large mammals in the Americas due to their unreasonable overhunting. Among them were horses but also other large herbivores, mainly woolly mammoths.

However, let's be clear: if the persistence of horse DNA in mid-Holocene sediments made Collin's model credible, it didn't fully validate it, since the proof showing that the horses of America had survived until the Age of Discovery (between 7,600 years ago and the fifteenth century) was still lacking. There remains a paleontological and molecular void of more than sixty centuries for the horses of the conquistadors and the hypothetical last survivors of Amerindian horses to have met. But Collin's model, as improbable as it might have appeared initially, suddenly could no longer be ignored, and deserved formal testing. My inclination went in that direction, and without me knowing it yet, it opened the door to a new project that would bring Collin and her family to France to work in my lab for close to three years. Here is my reply to the email she sent that August evening in 2018: "I should start by saying that nothing would surprise me anymore as far as the evolutionary history of the horse is concerned." At that point I did not have the American horse in mind but instead Przewalski's horses, which we had just shown not to be the last wild horses roaming on the planet, but to have descended from the first horses ever domesticated at Botai. What I added next was more to the point: "I would feel ashamed if I was trapped in propagating strictly Eurocentric views on horse domestication, and I realize the deep implications of your scenario. These deserve proper testing. I would thus be very keen to help you further test this using the best of my expertise in evolutionary genomics and ancient DNA research."

Backstage at Designing a Common Project

Collin soon visited my lab in Toulouse, and we felt an immediate connection. She told me about the existence of the Sacred Way Sanctuary, which she and her husband Sean dedicated to the preservation of Indigenous American horses. It brings together horses from a great many other Indigenous nations, such as the Lakota, the Cheyenne, the Cherokee, the Apache, and many others. The family history of some of those horses is well known and for some even goes back to the very beginning of the nineteenth century. By sequencing the genome of those horses, we hoped to be able to measure the portion of their genetic ancestry directly issued from the horses that once populated North America. We knew how it should look because Shapiro and I had already sequenced the genomes of horses that lived 28,000 to 35,000 years ago in what is today the Canadian Yukon. The plan was taking shape; we would measure the genetic portion that the horses of the Sacred Way Sanctuary shared with the latter, and would compare it to the one that is found in other horses, whether from America or elsewhere, domesticated or not, living and ancient. We would extend the sequence comparison to other horses of the Americas, notably the famous mustangs, and other populations such as the horses of Santa Cruz Island off the coast of California, which are known to have descended directly from the first colonial horses that the conquistadors brought to the New World.

If Collin's model was correct—that is, if the Indigenous horses and the mustangs were of mixed blood—we should right away be able to establish the genetic link with the nearly 30,000-year-old horses of the Canadian Yukon. We should also find that this link was stronger among mustangs and the horses of the Sacred Way Sanctuary than among other horses, in particular those whose ancestors had never set foot in the Americas. We were on a good path to formally test the model Collin was defending. Aware of the power of genetic tools and also of the weight and consequences of the results that we might obtain, I told Collin: if this is what we find, your model will be accepted by Western science; you will have revealed to us what was probably one of the greatest myths of the history of the horse, and perhaps of history itself. But if we don't find a direct genetic link, we will have to report that the version officially defended by Western academia is not a myth, and publish the results. She looked me in the eyes and said that she knew, and that the time had come. This was in fact one of the reasons why she came.

The plan was becoming concrete, but there were still gaps because the horses living today, even if they were descended from lineages going back to the nineteenth century, are not those that lived in the time of the conquistadors. Ideally, we needed to be able to sequence the DNA of horses from the sixteenth century and those of the first centuries of colonization. This is where Will Taylor, assistant professor at the University of Colorado, Boulder, came onto the stage. Taylor and I had collaborated for several years on the origin of horses of the steppe; we met for the first time in Mongolia to study the horses associated with deer stone-khirigsuur complexes, a type of monumental funerary site featuring sacrificial horses. This type of complex marks the appearance of an early, strongly hierarchized society in Mongolia during the Bronze Age that lasted until the beginning of the Iron Age, in which horses played a critical role. But that is another story. What concerns us here is a completely different project Taylor had brought me into when he was getting ready to go back to Boulder to establish his own laboratory: studying the archeology of horses and the equestrian practices in the Americas during the time of colonization. The timing was perfect.

Taylor hoped to undertake new archeological digs and study all the remains of horses that were scattered in various museum collections, investigating their morphology and scrutinizing the slightest clues indicating whether these horses had been ridden,

cared for, and fed—any sign revealing the nature of the relationship horses had with humans of the time. Together with his
collaborators, he also planned to subject the remains to delicate
chemical analyses informing horse mobility, which consisted of
measuring the presence of certain chemical elements (the isotopes of strontium, to be precise) in the teeth. Since these elements enter our bodies from the water we drink and the food
we eat, their presence in our tissue reflects the environment
where we lived at the moment they were ingested (a principle
we have seen with the Botai site in chapter 2). In this case it
would involve repeating the isotopic measurements along one
tooth, millimeter by millimeter, to evaluate whether the animal
had remained in the same environment throughout its life—
the amounts are stable along the tooth as it grows—or whether
it had moved and passed through other environments. The
icing on the cake: it happens that environments can have chemical levels that are very distinct from each other, so finding their
characteristic level in an animal is like having a tracer that enables one to locate precisely the environment it visited. With
these techniques at hand, we would perhaps be able to say
where the historical horses of the sixteenth century and those
that followed had lived, and at what point they had been mo-
bile. Taylor would also obtain an estimate of the date at which
the animals had lived by delicately calibrating the results issued
from the technique of radiocarbon dating. As for me, I would
provide the full support of my lab to Collin so she could sequence the DNA of those horses and have their genetic profile
analyzed. The definitive experimental plan was in place, and it
was solid. The historical model proposed by Collin could finally
be subjected to a rigorous test. All we had to do then was roll
up our sleeves.

Toward the True History of the Horses of the Americas

The genomes of 31 horses from the Sacred Way Sanctuary were sequenced, as well as those of 25 modern North American horses, including mustangs and horses from Santa Cruz Island; we also sequenced those of Hispanic horse breeds such as the Galicenos, who left Spain and reached Mexico in the sixteenth century, and the remains of 19 horses found among various archeological sites in Argentina and the plains of the American Southwest. The genetic verdict was pronounced: the type of horses to which every one of them was most closely related was unambiguously and always the same—it was the DOM2, descended from ancestors domesticated in the lower Volga-Don valley 4,200 years ago. As far as our data set could tell, the horses found in the Americas since the Age of Discovery all have a strange family resemblance not reminiscent of the horse that populated the Great American North in precolonial times; their genome is characteristic of the DOM2 lineage that began leaving the lower Don-Volga region some 4,200 years ago, and thus their principal origin was located in the Old World.

Our measuring instruments also indicated that the portion of the genome that may be related to another origin is scarcely detectable for some of the animals analyzed, and barely reaches 1 percent in others. These levels are comparable to what is sporadically found in some of the breeds of the Old World. This ultra-small fraction of the genome most likely reveals migrations following glacial ages, which we've already discussed, and left common genetic traces on either side of Beringia. It was not of the kind that would have been formed if the horse of the conquistadors had suddenly mixed with an Indigenous lineage that survived in the Americas until the sixteenth century. We had to face the fact: at first glance, the expectations of the model supported by Collin did not seem to hold up. At least for now.

But thinking a bit further: What if the historical horses we sampled did not capture the places where the Indigenous horse lineage survived? That remains a serious possibility: after all, no matter how careful we had been with our experimental design, we only analyzed nineteen archeological horse remains from the Americas, which does not look like an awful lot compared to the scale of the southwest plains and the continent. In a scenario where Indigenous horses survived at the time of colonization, we might also expect that their sacred remains would be protected by traditional knowledge-keepers and would not be readily available in the storage facilities of the Western institutions with which we collaborated. Furthermore, in thinking about the colonization and the numbers involved, it is worth asking how many horses were repeatedly imported from Europe, and how many of their descendants were bred in the Americas. It was likely a huge amount. Wouldn't this have rapidly ended up diluting to almost zero any contribution from the numerically less common Indigenous bloodline? More work is necessary and under way to test these scenarios, and to keep giving Collin's model the full attention it deserves.

The Šuŋkawakaŋ Nation

The data collected at that stage already conforms to another element of Collin's thesis, no less central in my opinion: the Eurocentrism of the official history of horses in the Americas. Rare are the direct historical sources that account for the Indigenous reality of the North American continent before the beginning of the nineteenth century. The official history largely contends that it was beginning with the rebellion of the Pueblo people in 1680, which was suppressed but put Spanish colonization on hold for a dozen years, that the horse began to reach the great plains of what today we call the southwestern United States and to multiply there in great numbers. According to that thesis, the Plains Indians would not have begun to establish a way of life centered on the use of the horse before that date. However, the radiocarbon dating Taylor obtained revealed the opposite: the remains of a horse found in Black's Fort in Wyoming, and those of another found in the Kaw River in Kansas, were from the second half of the sixteenth century and the beginning of the seventeenth. Another dating we had obtained in 2021 from a specimen from Idaho showed the same thing. The osteological traces indicate that the Indigenous peoples provided veterinary care to the horses, since some among the oldest we analyzed revealed that they had survived what should have been lethal cranial fractures. As for chemical analyses, they established that among those same horses some had been fed corn, a plant native to the Americas, and that during their lifetime they lived in a territory that extended well beyond the limits of the world then occupied by European colonists. These horses were clearly those of Indigenous peoples, who developed an intimate relationship with animals of European descent well before our history books tell us. Yvette Running Horse Collin, by collecting the stories from those who are never heard, had begun to reveal a part of history that had escaped Western academia: the precocious relationship that the Indigenous peoples had with the horse early on, to the point of recognizing in *šuŋkawakaŋ* the Lakota word for horse, a nation of its own.

And the story doesn't quite end there. The most ancient of the historical horse remains that we sequenced are genetically close to certain DOM2 horses in particular, and almost identical to those native to Iberia. It was only in the nineteenth century that the genetic profile of horses in America seems to have changed and acquired a closer resemblance to English breeds. This owes nothing to chance and reflects the great phases of colonization, first Hispanic, then British. The blood of American horses also bears the stigmata of colonization.

The complete history of this project is made up of many developments, all richly informative on both the human level and the scientific level. It would be untrue to say that it was always easy to cross-navigate continents and cultures that are built on fundamentally different foundations, starting with communicating in English, which is none of our native languages. For example, the concepts of "wild animal" and "domesticated animal" do not exist in the Lakota language, nor does that of extinction; beyond the horse, part of our work consisted of doing what was necessary to understand each other as equals, and to establish a strong methodological framework that would facilitate exchanges in a constructive way. It is that process, established in mutual patience, respect, and trust, that enabled us to go beyond our conditioning and the weight of history to construct a solid experimental process guaranteeing the acquisition of scientific data, and to accept the profound significance and current limitations of the results we obtained. I cannot help viewing this as the most important outcome of our work when reading Chief Joe American Horse stating in the original *Science* article published in 2023: "We stand with the horse and we will always do so however it has evolved through its journey. That is what being Lakota is." We therefore hope to have contributed to renewing the foundations of a constructive relationship with the peoples branded with a colonial past. Yvette Running Horse Collin ended all of her messages to me with "Mitakuye oyasin," which is most often translated as "In all friendship" or "With all my heart," but whose profound meaning I now understand: "In the name of all my relatives." Everything in nature is connected; us, our horses, and much more still.

11

The English Thoroughbred

The Fallen from the Racetrack

When he goes into the metal stall at the starting gate on May 20, 2006, Barbaro is clearly nervous. He has just suffered a first false start. The pressure is at its peak: this English Thoroughbred is getting ready to compete in the Preakness Stakes, the second of the three races that form the American Triple Crown and which on the calendar precedes the Belmont Stakes and follows the Kentucky Derby. Barbaro had just won the Derby two weeks earlier, and handily, by more than six lengths, a margin that hadn't been seen since 1946. If he managed to win again today, the horse would be on its way to claim the Triple Crown, a title that only eleven horses had won in close to a century of competition, joining a list of champions with unforgettable names whose prowess is still glorified today. Secretariat is one of the best horses that racetracks have ever known, and yet his time on each of the three tracks remains unbeaten in more than fifty years. In 2018, one of the shoes the champion wore the day of his victory at the Derby was sold at auction for more than $80,000. That just shows what a victory by Barbaro would mean.

This horse, barely three years old, was getting ready to race as the favorite on that May day. When the stall doors opened for the second time after the first false start, the close to 130,000 spectators around the racetrack—as well as millions of others who followed the event live on their TV screens—were all holding their breath. There wouldn't be a false start this time. The race had begun. In scarcely less than two minutes, if all went well, and after 1,900 meters, those who were betting would find out if their hopes of a nice profit would be fulfilled or dashed, and whether the animal, with its victory, would enter into legend.

The suspense would be short-lived, however, because the animal didn't even make it to the first bend. Very quickly it began to jump around. The competition continued their mad race at full speed, leaving the favorite far behind, alone. Barbaro could no longer put his right hind leg on the ground, and ultimately stopped moving, standing on three legs. Everyone immediately understood that the leg was broken. The veterinarian who took care of the animal in intensive care at the New Bolton Center hospital diagnosed more than twenty fractures. On January 29, 2007, eight months after its accident at the Preakness Stakes, the animal was euthanized following many post-op complications.

The Silent Slaughter

Unfortunately, Barbaro isn't an isolated case; rather, he's the tree hiding the forest. Let's leave the United States and go to Australia, a continent that is just as smitten with sprint racing and where every year the Melbourne Cup is held, a race which at 3,200 meters long is the most prestigious in the country.

With a first-place prize worth AUS\$4.4 million, and 220 million more at stake in parimutuel betting, the Melbourne Cup is one to "stop the entire nation," as they like to say there. That isn't nothing for a country where the financial volume associated with horse race betting reached almost AUS\$23 billion in 2010. Between 2013 and 2021, seven out of the over two hundred English Thoroughbreds who started this mythical race did not survive it. In 2014, for example, the stallion Admire Rakti, the betting favorite, died of cardiac arrhythmia a few minutes after crossing the finish line in last place. He had just gotten back to the stalls. The stallion Araldo, who had finished the same race in seventh place, broke a leg as he was returning to the stalls and couldn't be saved, either. The fracture that in 2020 struck down Antoine Van Dyck, the stallion with the name of a famous seventeenth-century Dutch painter, in the middle of a race ended the same way; he soon had to be euthanized. Beyond the Melbourne Cup, global statistics for the country speak for themselves: not even half of the horses that begin on racetracks at the age of two or three—barely 46 percent—will still be active two years later. The losses are therefore on par with the financial stakes—in other words, considerable.

A horse's body isn't fully developed until its sixth year, but even if the risks of injury from fractures or the results of torn muscles or tendons are enormous, many begin their training scarcely eighteen months after they're born. This would be the equivalent of making children barely ten years old enter professional competitions. It is true that the low life expectancy of racehorses does not statistically encourage taking one's time, especially given the astronomical winnings that are sometimes at play in races reserved for two-year-old foals—such as AUS$2 million for the Magic Millions race. All of this feeds the hope for a rapid return on one's investment on the financial level, and often ends up convincing owners to line up their budding champions at the starting gate. The hashtags #enoughisenough, #youbettheydie, and #nuptothe-cup have flourished and accompany a galloping awareness in Australian society of the heavy cost to the animals on the race-track, but little seems to have changed; an extra 4 billion Australian dollars were added to the financial till associated with horse betting between 2010 and 2018, reaching close to AUS$27 billion.

Allegations of Doping Today

Let's go back to the United States for a moment, and to the Kentucky Derby, Barbaro's last victory. Medina Spirit, a three-year-old English Thoroughbred, finished that race as the winner in 2021. His victory remains clouded in controversy, however, because the stallion was also declared positive during the antidoping test that was done right after the race. Despite his trainer, Bob Baffert, claiming that the banned steroid detected came from a treatment against skin inflammation that should have no effect on the race outcome, Medina Spirit would become the third winner in 147 years of the race history to be disqualified. The animal passed all additional testing to enter the Preakness Stakes, in which he finished third. However, six months later, he didn't survive a final warm-up session just days before the Santa Anita race, collapsing with no evidence of doping, possibly due to a heart attack. If the affair shook the California racing milieu, it was not just because it was yet another premature death of an exceptional horse; it was also because it could shed an icy light on the entire business.

Medina Spirit was not the only horse trained by Baffert whose urine revealed illicit substances. Justify, another of his champions, winner of the Triple Crown in 2018, had also tested positive. There are in fact at least thirty horses that failed drug tests. According to the *Washington Post*, which listed the trainers who had lost at least one of their horses since 2000 in California, there were no fewer than seventy-four horses who had died under Baffert's training. With six victories at the Kentucky Derby, eight at the Preakness Stakes, three at the Belmont Stakes, and being the trainer of the only two horses who had taken the Triple Crown since 1978, the man displayed one of the most prestigious resumes in the history of horse racing. The Baffert method incontestably produced champions and brought in more than $320 million dollars in winnings. But the sport as a whole was also fatal, it seems, to a considerable number of contenders for the top racing prizes: combined, the first ten trainers at the top of the list published by the Washington Post have lost 617 horses since 2000.

The story of Medina Spirit brings together all the ingredients of a feel-good story: the stallion had cost only $35,000 when he was bought, at a time when the seeds of champions were easily worth hundreds of thousands, and even more. He was not the top favorite when he set out for the Kentucky Derby, the reigning race of its kind in the United States. His odds were 12 to 1. Only the anti-doping test that revealed the presence of betamethasone, the corticosteroid known to reduce pain and swelling of the joints, began to cast a shadow— until his death in Santa Anita. Years earlier, the sudden deaths of seven other horses trained by Baffert already triggered an investigation, which revealed the administration of thyroxine—a hormone naturally produced by the thyroid gland that gives a metabolic boost to the organism. The trainer was cleared by the California Horse Racing Board as there was nothing illegal about using this treatment. Today we know that an overdose of thyroxine can diminish racing performance and cause sometimes fatal heart arrhythmia.

Doping in the Past

From the time of Al Capone, and up until the 1970s, it wasn't uncommon for cocaine to be given to horses to enhance their performance. The effect was sometimes the opposite of the one anticipated, but other methods could be used to facilitate the victory of a young horse. Arsenic poisoning could be used to eliminate the most serious competition—recall the case of Phar Lap in chapter 2. Less radical, but just as cruel, was stuffing the back of a horse's nostrils with sponges to greatly reduce its ability to breathe, thus also reducing its optimal speed. Since that time the arsenal of drugs with doping effects has grown, and their use has become more sophisticated. The use of cocktails of several molecules expertly dosed to remain under the legally allowed thresholds appears to be one of the very popular strategies.

The furosemide-phenylbutazone combination provides one such example; the first molecule is reputed for reducing the risks of bleeding in the lungs during intensive exercise while the second is an anti-inflammatory that stimulates the organism and makes it forget the sensation of fatigue. Banned in Europe, furosemide remains authorized in the United States. Another product banned in horse races worldwide is scopolamine, whose effects dilate the bronchial tubes and can improve horses' respiration and cardiac rhythm.

The situation is gradually changing because certain race-tracks, notoriously lethal for horses, have been closed while official inquiries attempt to shed light on the circumstances surrounding inexplicable sudden deaths or accidents that lead to the euthanizing of horses. This is the case in Santa Anita, where a policy of zero tolerance is in place today. In 2019, the governor of California imposed a halt on racing at the height of the season when twenty-nine racehorses died there. It nevertheless remains that out of every 2,000 horses that will go to the starting gate for a race in the United States, an average of three will end up dead. That number climbs to seven in California, or close to six times more than in Hong Kong and almost ten times more than in Great Britain, which means that every year several hundred premature deaths can be attributed to races organized on the American continent alone.

The use of drugs might be a factor in such numbers. Some, like scopolamine, which enhances horses' respiratory capacity, can also lead to pulmonary hemorrhaging. Similarly, one synthetic hormone marketed by the Epogen company, though banned in the sporting world, is theoretically prescribed to treat anemia; it stimulates the production of red blood cells and thickens the blood, which again enhances respiratory capacity but can have serious side effects if misused, such as heart attacks or cerebral and pulmonary embolisms. Rumor even has it that some trainers have used snake venom because of its anti-pain properties. A horse doped in that way could no doubt surpass its own abilities but might also be pushed to the point of collapse, not being able to feel the warning signals his body is sending him.

Fragile Bodies Pushed to the Limit

Despite the boost that chemical substances may provide, in English Thoroughbreds death most often occurs following fractures sustained in training or during a race. Those horses, often too young and overtrained, are not allowed the time they need to recover and rest. When a bone in the leg fractures at speeds close to 70 kilometers per hour, the effect is often fatal for the animal. Its legs are particularly delicate, and lacking tissue able to protect the bones, fractures often break through the skin and quickly lead to infection. It isn't uncommon for fractures to cut the blood vessels and stop blood circulation to the stricken limb. Even when such complications are avoided and fractures are successfully reduced, a serious difficulty further prevents the road to recovery: the weight of the horse itself, which reaches close to a half-ton on average. It is difficult for a racehorse to support its weight on three legs for several weeks. A chronic imbalance can settle in, and with it founder, which is known by the scientific name of diffuse aseptic pododermatitis and which is enemy no. 1 of all breeders. This is an acute inflammation of the foot that is difficult to treat, whose symptoms Xenophon had already described around 400 BCE and which at worst results in the perforation of the hoof by the bone of the last phalange and ultimately in the animal's death.

Reproduction Under Close Surveillance

English Thoroughbreds are at the heart of an industry that earns close to $30 billion in annual revenue in the United States alone and employs almost a half-million people, including those in veterinary clinics, stables, and training centers and racetracks. If the athletic performance of these animals is the essence of the system, their reproduction is no less so insofar as it concerns producing both prodigies of speed generation after generation, and also a sufficient number of foals every year—around 100,000 worldwide—to continue to satisfy the demands of a globalized industry. An animal's reproduction is not left to chance. Far from it: it is standardized, optimized, and meant for profit.

A foal's birth date is truly a high-stakes matter. In the Northern Hemisphere, by agreement all foals, whatever the true date of their birth may be, are given January 1 of the year they are born as their birthday. The practice aims to facilitate the organization of races by arranging for horses of the same cohort to compete against each other: that is, two-year-olds against two-year-olds, three-year-olds against three-year-olds, and so forth. This means that foals born on New Year's Day, Easter, or Christmas in the same year will race against each other. It doesn't matter that one has lived almost a year more than another, since their official registers will show the same age. However, there is no doubt that when foals officially enter competition at the age of two, the size of a January foal and his chances of winning will be much greater than those born in December.

As we can see, competition for an English Thoroughbred doesn't begin on the racetrack, but at conception; this means that breeders put conditions into place so that a pregnancy begins and ends as early as possible in the reproductive season. The same trick is also found in the Southern Hemisphere, the only difference being that the standardized birth date is fixed at August 1 due to the inversion of the seasons.

Being able to play with the calendar this way assumes following a well-oiled reproduction technique. Future mothers are kept inside in the winter and exposed to artificial light for periods of time that increase slightly day after day, a ruse that suggests spring and its lengthening days. The hormonal cycles connected to reproduction are put into place earlier in the year, and the female will then ovulate earlier. Synthetic hormone treatments are able to help if the natural mechanics are delayed.

A mare's menstrual cycle lasts twenty-one days on average. It includes only seven days during which she is receptive to the advances of a stallion, so it is indispensable for a breeder to present to her at the most opportune moment the stallion he has chosen in the hope of obtaining a future star on the racetrack. It is also just as crucial that the act of copulation results in fertilization on the first try: each covering costs money, and if the mare is not inseminated, there will be a delay on the calendar of optimal births. Ovulation is followed very closely by veterinarians, often with the help of daily rectal ultrasounds.

One must also ensure that the mare is in the best mindset to be covered. Care is taken to present her with a substitute stallion: a sort of gigolo who will court her, as if to put her in the mood, and most often a pony much smaller than she, because he will then not try mounting her. He will be removed at the right time to leave room for the breeding stallion, but before that, the mare will be taken to her new quarters to prepare her for the act. Her genitals will be washed, her tail will be bound so the stallion's precious penis won't be harmed, and she will be draped with a covering aiming both to protect her from the biting that sometimes accompanies this type of activity and to hamper her legs, so she will become resigned to the maneuver and won't try to kick. It also happens that her upper lip is bound, because that practice seems to reduce her stress level, at least temporarily.

That is when the true stallion will make his entrance. While a mare undergoes the process only once, the breeding stallion will be able to resume his services immediately afterward. Some of them will have a very busy season, such as the Australian stallion Encosta De Lago, who is said to have covered the astounding figure of 227 mares in 2008. It is due to their prestigious genealogy but also to their own wins that the service of stallions can go for the price of gold, reaching $1 million per covering for Northern Dancer, for example. This Canadian stallion was named best foal of the country in 1963 and best horse overall the following year, and was the winner of the Kentucky Derby and the Preakness Stakes as well as twelve other races, earning him the tidy sum of more than a half-million dollars in race winnings—an almost paltry sum compared to his earnings in his second career as a stud. Putting the most prized stallions out to stud can earn their owners a lot of money.

The stakes are such that the most novice among the horses receive special training for it, and sometimes assistants intervene to guide penetration with instruments created especially for that purpose. Two weeks after covering an ultrasound will verify the beginning of gestation in a mare, without which she will be sent back to the task during the following menstrual cycle. If the ultrasound reveals the presence of twins, one of the embryos will be extracted to limit the risk of miscarriage; if everything goes well, eleven months later, an entire team will be in attendance to help her bring her foal into the world. If it happens that the young mother is too weak to take care of her own foal, it can be given to a wet nurse who will take care of it like her own. In any case, barely a month will have gone by before the infernal cycle may start all over again, because with eleven months for gestation, the new year approaches quickly.

Life as a Lottery

It is estimated that around two-thirds of all Thoroughbred foals born will follow the path to the racetrack; some will be wounded before they can race, others will have the life of breeding horses, or will quite simply be judged unfit. Those who are born under a lucky star will be able to be sold at exorbitant prices in their second year, like Snaafi Dancer in 1983, one of the many sons of the prodigious Northern Dancer, who was the first yearling in history to exceed $10 million at auction. In an irony of fate he would end up participating in no official race, as his training time was much too slow to hope he would have the slightest chance of victory in competition. He had hardly any more success as a breeding stallion, as he had fertility problems. Betting on the genetic lottery means that you can win or lose, despite all the attention bestowed on the process of obtaining the best chances of winning.

Overall, the life of an English Thoroughbred is far from the long, peaceful journey one might imagine if one remains blinded by a romantic passion for this incredibly elegant animal. It can prove to be truly difficult—as we have seen—beginning at the stage of its conception, and continuing in its training, on racetracks, or in stables. And we mustn't imagine that once taken off the racetrack these horses will necessarily enjoy happy days and have a well-deserved retirement. This is true for a large portion of them, and so much the better. But others finish their days at the slaughterhouse, as industrial meat destined to feed our pets. This even happens to some of the most prestigious among them.

This is how Ferdinand, winner of the Kentucky Derby in 1986, twenty years before Barbaro, ended up. No one would have predicted this fatal destiny for the horse that had been crowned the best horse of the year and whose cumulative wins brought in close to $4 million in purses. He was taken off the racetrack three years after his Derby win, and at the time it cost $30,000 to have him cover a mare. However, his offspring did not distinguish themselves with particularly brilliant performances on the racetrack. The price for him to copulate fell rather quickly, so that soon the animal was no longer living up to the hopes his owners had for him; he was sold to a Japanese breeding center, where he spent the next six years, which were his last. He initially had some success there, covering seventy-seven mares in the first season, but ultimately people lost interest in him again. He covered only ten during his last season. Negotiations to place him in a riding club failed, and at the beginning of 2001 he fell into the hands of a horse merchant before ending up the following year in a slaughterhouse in the Land of the Rising Sun.

Ferdinand's story is tragic, without a doubt. It nevertheless had the upside of casting a light on practices as questionable as they are revolting, and which remain all too often the fate of many horses, most of them losers and removed from the racetrack. In Australia, the third industry for English Thoroughbred racing in the world, a protective law bans horses from being euthanized without a veterinarian certifying the medical merit of such a decision. But that law isn't applied universally unless the animals are still performing. Once their racing career is over, less stringent rules are applied, depending on the region, and seal their fate. A large number of horses still end their days in the slaughterhouses of Australian states like Queensland, where there is a race to the bottom of the judicial scale that aims to ensure animal well-being.

Reasons for Hope

The facts would be depressing if we left them at that—so we must at this point give credit to the people who work to improve the conditions of horses. If the lure of profit can push some people to do whatever they can so their horses cross the finish line first, a love for this animal prevents others from resigning themselves to its fate. There is increasing action to push such boundaries. There are, for example, social networks and hashtags such as #youbettheydie and #enoughisenough that contribute to informing our societies about practices that are too often carried out in silence. There is also the Ferdinand Foundation, which, on the initiative of the NYRA (New York Racing Association), has adopted the mission of financially supporting the care of horses after they leave the racetrack. In addition, there are an increasing number of refuges and convalescence centers that, following the example of the Kentucky Equine Adoption Center, have for fifteen years offered a peaceful haven in the heart of the American state most influential in breeding racehorses, and not just for English Thoroughbreds and the other bloodlines valued by the racing industry. The refuge has taken in more than a thousand horses in distress, along with those that have been judged unfit and abandoned. Other initiatives, just as positive, seek to push the limits of science and veterinary medicine to contribute in their way to this battle.

Let's take for example the new therapeutic approaches aiming to care for tendon wounds, which along with the suffering they cause, often put an end to the racing careers of champions. In 2012, the stallion I'll Have Another was on track to win the American Triple Crown, an exploit that no other horse had achieved since 1978. Having won the Kentucky Derby and the Preakness Stakes, he just had to win at Belmont to take the coveted title. But a simple tendon wound was enough to stop him and keep him from racing for the rest of his days. He finished his career as a breeding stallion.

This is because tendons, like our Achilles tendon, which bear the weight of the body, are subjected to massive physical stress during exercise. They don't just connect muscles to their bony support; they function as true springs that absorb energy during contractions to distribute it during extensions. That principle, combined with a particularly elastic spine, contributes to making a horse's gallop one of the least energy-consuming movements in the animal kingdom. Those of us who follow the careers of high-level athletes know that a ruptured tendon can lead to a convalescence of several months. Horses don't escape that rule, because it involves tissue that has a relatively small blood supply, is low in cells, and is characterized by slow biological healing processes. It can also happen that a wounded tendon doesn't entirely recover either its resistance or its flexibility even once it is healed. This is a serious problem if horses go back to competition, for their well-being and for their performance as well as the higher risk that their injury will recur.

Toward Cellular Therapies

Innovative cell therapies are emerging involving stem cells, which are injected directly into the wound site. The stem cells used have the same properties as the first cells that divide during embryonic development; beyond the power to divide, they also have the potential to ultimately become any of the cells in our body, including those that make up tendons and manufacture their characteristic collagen fibers. We would face an unsolvable ethical problem if we could only obtain these stem cells from embryos, because it would then be necessary to end the development of a donor individual with the single goal of hoping to heal the wounds of another. The good news is that it is possible to derive these stem cells from tissue of adult individuals. (This is one of the great discoveries of the early twenty-first century, which in 2012 earned the Japanese scientist Shinya Yamanaka the Nobel Prize in Medicine.) To do this, one must manipulate these cells to induce their manufacture—a cellular biologist would say their expression—of several proteins that bear the name of their discoverer: Yamanaka factors. The resulting stem cells are called induced pluripotent stem cells, to point out that there was manipulation and to distinguish them from embryonic stem cells.

In principle, a simple biopsy taken from a horse at the top of its form—most often from the skin, fat, or bone marrow—is enough to isolate, produce, and maintain the stem cells in the lab. They can then be used in the case of injury during the animal's entire career (this is called autogenic injection). One might even imagine injecting these cells into another individual (allogenic injection), on the condition that one can suppress the expression of other factors that would signal to the host's immune system the intrusion of foreign cells and would trigger an immune reaction leading to a local inflammation that could impede healing. Trials in this area are being undertaken, and in the past few years have shown encouraging results in avoiding both inflammation and the formation of teratomas, which are uncontrolled divisions of cells that form a sort of tumor at the injection site.

As for autogenic injection, the first clinical trials were carried out in the middle of the 2000s. These trials demonstrated a better healing prognosis and better chances of returning to the racetrack twelve months after an injury. More recent studies have confirmed a clear tendency toward a better organization of the collagen fibers of the tendons post-treatment, and faster progress in healing. In high-jumping competition, the rate of regression two years after treatment appears reduced, and most often the horses regain the level of performance they had before the injury. Buoyed by these results, equivalent strategies are in the testing phase to fight against other types of pathologies linked to an intensive practice of the sport, including muscle tears and the wearing down of joint cartilage, or even the dramatic founder mentioned earlier. They are based on the ability of stem cells to multiply locally and to become all types of cells the tissue needs to heal, as well as the ability to emit chemical signals that recruit other cells that may encourage healing.

The Arms Race Through Genetic Doping

The combination of high-level sport and cutting-edge technologies is also found in a new form of doping that has exploded on the racetrack. While previously only chemical, doping today is crossing a new boundary and becoming increasingly genic—transgenic, to be exact, since it is much more than just simple molecules that are injected into the animals. Highly sophisticated genetic constructions are used—DNA fragments or vectors, which are organized in such a way as to produce substances with a doping effect by the cells of the organism itself. For example, there is erythropoietin, that hormone better known as EPO, which has often seduced Tour de France cyclists because it stimulates the production of red blood cells to increase the capacity to transport oxygen in the blood. There are others too, and with a simple injection of this new type of vector the promoters of this research hope to stimulate the secretion of a true cocktail of products, each furthering performance gain. A very recent study has listed close to thirty vectors in addition to EPO: these are the ones known today to stimulate racing performances, and there may well be more in the future as the advancement of our knowledge in this area increases.

Several technologies have already been proven to detect the presence of these vectors in the blood and urine of champions. In general, they rest on techniques of amplification surgically targeting through PCR (polymerase chain reaction) genes with doping effects as they are organized in the vectors. That way the doping agent will be not confused with genes as they are in their natural state, in the genome of the animal. Though these techniques are extremely sensitive and can detect the handful of copies that might be hidden in barely a drop of fluid, they suffer from two big limitations at this stage. Let's not forget that Lance Armstrong, six-time winner of the Tour de France, never tested positive for EPO even though he was doping, as he subsequently admitted.

It's important to keep in mind that the injected vectors are not integrated into the DNA carried by each of our cells, and end up being naturally eliminated by the organism; they are only detectable within a very short window of time, typically in the three days following an injection. But the effect of doping products can last or even develop over a much longer time frame. That is the whole problem: as long as random and repeated controls are not put into place, one can calculate a plan that will pass the net of discovery.

The amplification techniques for these transitory transgenes also make it necessary to know their target—the famous genetic construction I mentioned earlier. Their content (genes), as well as their possibility for arranging (the order and organization of the genes), are almost infinite. Defrauders have only to use a little imagination to avoid surveillance; their system will always be ahead of those meant to detect them. We must therefore fear that the race-pursuit launched in molecular biology labs is often won by the defrauders even before it unfolds in racing arenas—unless the total DNA sequencing extracted from blood plasma or urine enables the detection of an unusual presence of genes that do not completely resemble the chromosomic versions carried by the individual. Proof-of-principle studies exploring the potential of total DNA sequencing seem to indicate that the race is perhaps not lost in advance.

The Quest for Performance

As we can see, the history of the English Thoroughbred has essentially been guided by a single obsession since its beginnings: the quest for performance. At the beginning of the seventeenth century King Charles II of England, a fan of racing and the founding father of the horse racing temple at Newmarket, near London, laid the foundations of the sport: races could be run over longer distances than they are today, at least 3,000 meters and sometimes even over 6,000. That is much more than the standard 2,400 meters that has dominated since the middle of the nineteenth century in sprint racing among English Thoroughbreds (other horses, like the American Quarter Horse, compete over much shorter distances). Breeders have been able to do their selection work for more than 150 years with just one goal in mind: to obtain two- to three-year-old foals that are not especially endurant, because over such distances a second wind is a less important quality than being capable of the highest possible top speed.

However, for the past sixty years, while this work of selection continues and the wins of horses determine how bankable they are for breeding programs, the records of the most prominent races have only stagnated. Citing only those of the American Triple Crown, they are still held by Secretariat and his legendary races of 1973—leading some to think that the sport, which requires maintaining average speeds of around 60 kilometers per hour for more than two minutes, has reached the absolute limits of what the biology of the horse can achieve. Others believe quite the opposite and explain the stagnation by asserting that the horse isn't really fighting against the clock but against other horses, and that psychological factors or the racing strategies adopted by jockeys, rather than the mechanics and physiology of movement, are to blame.

Myostatin and Sprint Genetics

One thing is certain: the selection work by breeders has changed the genetic composition of Thoroughbreds generation after generation. This is certainly the case for isolated genes such as the *MSTN* gene, which codes for myostatin and has earned the nickname as the "gene of speed." In 2010 three research teams discovered that there were two versions of the gene on horses' chromosome 18, and that a difference of only one letter among the more than 80 million of that chromosome exerted a major influence over sprint performance. If a horse's father and mother both transmitted a first version of this gene (the one carrying the letter C), its sprint speed would on average be faster than if they had transmitted only the second version (the one carrying the letter T). A horse carrying the two versions at the same time would demonstrate intermediate performances on average. It was discovered that this gene acted like a switch whose first position leads to a maximal top speed, the second to an intermediate speed, and the third to endurance abilities instead of speed. Among the elite of the English Thoroughbreds, it appears logical to find more than two-thirds of horses at the first level, and none at the last. The first have inherited at birth a genetic patrimony conferring more chances of winning victories in sprint races; all things being equal, their performance is better overall and consequently breeders will tend to choose them more often in their breeding programs. Generation after generation, therefore, the version of the gene they carry—letter C—was able to invade the population. This is the principle behind artificial selection.

Later work has shown that the true causal mutation for increased performance was not the one we thought but another located not far away, which is transmitted almost always from a block with the letter C. It corresponds to the insertion of a DNA fragment of viral origin 227 letters long whose effect is to reduce the level of expression of the *MSTN* gene. The cells that carry these 227 letters manufacture less myostatin, a protein that limits muscle development, with the result that horses that carry this mutation develop an enhanced musculature and sprint capacity.

It has long been thought that English Thoroughbreds owed their speed to their Eastern origins. The Thoroughbred register is one of the oldest of its kind in the world and records all the crossings that have given birth to these horses since 1721— and three Eastern stallions have exercised an absolutely overwhelming influence. Their names are known by all race enthusiasts: Byerley Turk, who was captured by the English in 1687 following the Ottoman siege of Buda in Hungary; Godolphin Arabian, who was imported from Yemen before being gifted to Louis XV in 1730 as a diplomatic gesture, and then going to England; and Darley Arabian, about whom we know much less, except that he was imported in 1704 from Syria, probably from the environs of Palmyra or Aleppo. For once, a genetic analysis looking not at horses living today, but at ancient horses, would deflate that legend.

The remains of twelve stallions who were born between 1764 and 1930 are located in various natural history collections, including those of the Museum of Natural History and the Royal Veterinary College, both in London, and the University Museum of Zoology in Cambridge. Among them was Eclipse, a legendary horse unbeaten in eighteen races and great-great-grandson of Darley Arabian; collectively, these twelve specimens descended in a straight line from the three founding stallions. To everyone's surprise, DNA sequencing revealed that none of them carried the version of the *MSTN* gene containing the letter C that confers an aptitude for sprinting. It was difficult to imagine that their illustrious ancestors could have carried it without having transmitted it to them. The evidence spoke for itself: it was very unlikely that the speed gene, in its version associated with sprinting, could have the exotic origins the legend claimed. Statistically, it seemed much more likely that it had been transmitted by one of the English mares of local extraction and that it had crossed with one of the Oriental horses, or one of its descendants. The analysis of genetic genealogies indicated that the sprint version of the gene had remained relatively rare in the eighteenth and nineteenth centuries, but that enthusiasm for selecting for it didn't really take off until the 1960s, under the influence of a very specific stallion. That stallion, born in 1961 and already mentioned here for his coverings that could bring in a million dollars, was Northern Dancer.

That a champion on the level of Eclipse doesn't carry the mutation of the *MSTN* gene associated with sprinting might seem paradoxical but it is not, for several reasons. First, recall that in the quarter-century that preceded the French Revolution racetracks were much longer; one of the races Eclipse ran in Winchester on June 13, 1769 was around 4 miles long, or close to 6.5 kilometers. Eclipse won, despite a handicap of 86 kilos; it was good that endurance also had a role to play. This suggests that it isn't enough to have just one mutation for a horse to be a star on the racetrack; if that were the case, any not-even-clever geneticist would have no trouble predicting the results of races. The exceptional qualities that make up the seeds of champions are the fruit not of one gene but the combined effects of several—probably a lot—that we are just beginning to discover. In addition to *MSTN*, there is the *COX4I2* gene; here too a difference of a letter is overrepresented in sprinters (in this case the T and not the C), and the product of the gene does not intervene in muscle mass but in the chain of reactions underlying respiration in cells. It is believed that it might influence metabolism during effort. There is also the *PDK4* gene, carrying the letter A rather than the letter G at a very precise position, which acts on the metabolism of glucose and is translated into favorable effects in sprinting.

The exceptional qualities that constitute champions also depend on their training, the care they receive, and on the entire environment that surrounds them, including doping. The inheritability of racing performances—the time taken to cross the finish line, for example—shows values much lower than the 100 percent expected if only genes counted. According to estimates, it oscillates between a few percentage points and around thirty percentage points at most. The environment in the broad sense plays a preponderant role, and this is why races have that very special ability, that electric charm, and that such inimitable art of always being able to thwart our expectations.

The 2018 edition of the Kentucky Derby had already provided the proof of this. As noted earlier it was won by Justify;

Justify's father was the American Thoroughbred Scat Daddy, and three of his other sons were also at the starting gate. Even as half-brothers, they still remained very close genetically. But although Justify ended up winning the race, Mendelssohn, one of his half-brothers, finished 20th, the very last. With morning line odds of 5 to 1, he had been the second favorite among betters, right behind Justify. The other brothers finished in the middle and third-to-last.

The Risks of Galloping Consanguinity

The anecdote might be funny if it didn't have consequences on the genetic level. At the 2018 Derby, three other stallions who started also had the same father, Curtin, and an additional three were also brothers, the offspring of Medaglio d'Oro. The genealogy of each of the twenty horses competing in the race that year inevitably went back to the same Mister Prospector, a stallion who was born in 1970, the same year as Secretariat. Even when they weren't brothers, in other words, the champions of the bunch were more or less close cousins.

The problem is not the levels of kinship in itself; the risk that I run of developing a certain genetic condition or another doesn't depend on the number of my aunts and uncles, nor of my brothers and sisters. It depends only on my parents, and more precisely on whether or not they are carriers of a causal mutation. On the scale of a population, if that mutation appeared on average once in every 1,000 individuals, its frequency would be one out of 2,000, because we possess two copies of every gene. When my parents are not related, the risk that I carry that mutation in two copies would be 25 out of 100 million. In other words, the risk of developing that condition

THE ENGLISH THOROUGHBRED   195

would be extremely low, and in a population like the French population, only a small number of individuals would be likely to be afflicted with that condition.

However, if my paternal grandfather were also the father of my mother, the situation would be quite different. My father and my mother would be half-brother and half-sister; the risk that their father (my one and only grandfather in our example) has transmitted the mutation to each of them is one out of four (identical to that of getting two "heads" in a row when flipping a coin). The risk that they would transmit it to me would be no different. In the end, the risk that I carry the mutation in two copies is one chance in 16,000; if I had been the result of incest, the risk that I develop the condition would be 250 times greater than through the simple effect of chance. My chances of surviving and reproducing—my "fitness," as Darwin would have said—would very likely be affected by a serious handicap if I had been the result of such a consanguine union.

Of course there are less extreme forms of consanguinity; unions can also involve more or less related cousins. However, the principle is solid. Incest and consanguine crossing are not taboo among breeders of English Thoroughbreds, and the importance granted to the wins of individuals and their progeny in the choice of genitors automatically leads to repeated unions between related horses. The logic guiding the entire industry thus functions as a powerful generator of consanguinity. The risk that horses end up developing genetic conditions is increased significantly, and consequently also the risk that their health, their life expectancy, and their fertility will be diminished. Recent studies, based on results obtained from thousands of horses, have estimated that the racing performance levels of English Thoroughbreds diminished proportionately to their level of consanguinity. It has even been estimated that an English Thoroughbred loses 7 percent of its chances of one day taking the path to the racetrack each time its consanguinity increases by 10 percent. In other words, a horse's immediate genealogy in part conditions its career even before it is born.

In reality, the effects linked to inbreeding depression have every chance of being even more serious than our example suggests. The founding population of English Thoroughbreds isn't counted in thousands of individuals but in scarcely a few dozen, according to the tallies made possible by their official register. If only one of them carried a harmful mutation, its frequency would be not 1 out of 2,000 in the beginning population as we imagined it, but in the range of 1 percent. In these conditions, the risk that a nonrelated crossing would produce an individual carrying two copies of the mutation is 1 out of 10,000 (rather than 25 out of 100 million), and remains 250 times greater in the case of incest. When we know that only ten individuals alone are enough to explain more than 80 percent of the consanguinity level present in the Australian population of English Thoroughbreds, we immediately see the nature of the problem. What does it matter if they develop pathologies later in their life once their athletic prowess is depleted, provided that their performances between the ages of two and four—the ultimate criterion of selection—are what they should be? Their harmful mutations then have all the space needed to be propagated in the following generations, even when they might be lethal for the embryos that might carry two copies of them. Genealogies and the history of the crossings they trace have not prevented some of these mutations, such as the one involving the *LY49B* gene, which is expressed very early in development, from reaching frequencies close to 20 percent in Japan and Australia.

From the Greeks to the Present Day

If the first Greek Olympiads, more than seven centuries before our era, already included sprint races, and if that sport was able to give birth to animals as sensational and magnificent as the English Thoroughbred, we must never forget the heavy toll our passion for racing has exacted and continues to exact on the animals today. The contribution of the tools of genomics is not limited to shedding light on a reality that is sometimes difficult to look at. These tools can also provide solutions to improve the current situation. We have seen this when they contributed to restoring the true history of this legendary animal, or when they aim to thwart emerging forms of doping, one of the inherent scourges of the industry. By characterizing regions of the world where exploitable cradles of genetic diversity still survive, these tools offer the hope of helping to remedy two other problems that undermine the industry that have only gotten worse in the last fifty years, due to the preponderance of a few favored genitors: galloping consanguinity and a constant erosion of genetic resources.

If the industry responds to the same logic in the United States, England, Australia, and South Africa, the influence of certain champions has not disseminated homogeneously everywhere, for reasons sometimes linked to history—in response to Prohibition in the United States, for example, England refused to register horses issued from American parents from 1913 to 1949—but also due to costs and geographic proximity. If Northern Dancer remains one of the most prolific stallions of the twentieth century, his lineage has especially penetrated Australia and Europe. In the United States, other stallions not directly related to him have also had top billing. This is the case with A. P. Indy, the grandson through his mother of

Secretariat, who has remained one of the ten preferred genitors of Americans for an entire decade. His descendance might advantageously reinforce genetic diversity outside the United States. That is the principle. But if genealogies can guide breeders in their choices by enabling the quantification of levels of consanguinity of each horse, the tools of genomics have much more to offer; they can use more than 2.5 billion letters to characterize one by one the mutations carried by every animal, those that are harmful as well as those that are beneficial. They carry the seeds of many precious clues that might help breeders make the best decisions possible to protect their investment and guarantee the long-term survival of their industry—and through that also prevent, we hope, much suffering by the horses.

12

The Horse of the Future

Polo: A Sport at the Forefront
of Animal Cloning

At Kheiron Biotech, a laboratory in the suburbs of Buenos Aires, the animal that appeared on the double glass doors on the second floor of the building immediately caught my attention. It had a strange appearance: four legs and a tail but no head, and where its neck should have been there was the upright bust of a man, with his fist raised. A centaur. But could I have expected anything less than to cross paths with a phantasmagorical animal when I was about to enter, that day of December 2018, into the heart of the laboratory known, for close to fifteen years, for being involved in the horse trade of a very specific type—clones? And not just any clones: Kheiron Biotech works with horse clones chosen to be included among the best in the history of polo, a sport that in Argentina is as famous as soccer.

Kheiron Biotech is not the first company to have achieved the feat of equid cloning. The scientific world had already seen the successful cloning of a mare in 2003, and a male mule, a usually sterile animal. The Crestview Genetics company in Texas made headlines in 2009 when it cloned a stallion by the name of Aiken Cura, considered at the time to be the best horse that polo player Adolfo Cambiaso had ever ridden. That's significant when one knows that every year for more than two decades, Cambiaso has earned the title of the best polo player in the world. With unequaled wins, the forty-something Argentinian remains the living legend of the sport.

When Aiken Cura broke his left front leg in a violent melee of mallets during the last chukker of the Argentine Open finale of December 2006, Cambiaso immediately jumped off his mount to try to ease his pain. He begged veterinarians to save the animal, and made sure every surgical operation possible and imaginable was undertaken, even amputation and the placement of a prothesis, but to no avail; he ultimately had to have the animal euthanized two months later to stop its suffering. Even though Cambiaso and his team, La Dolfina, ended up winning the tournament, and Aiken Cura was designated the best horse in the finale, the victory would forever leave a bitter taste in everyone's mouth.

Cambiaso's attachment to this animal was such that he could not resign himself to seeing it forgotten; he swore he would do everything in his power so that his story, and his influence on generations of horses to come, wouldn't stop there. The accident struck down the animal at the age of eleven, in the prime of its form, and beyond its performances in competitions it had been destined to leave behind a prolific descendance of little Aiken Curas who would have been able to perpetuate his lineage. Cambiaso was contacted by Alan Meeker, an eager Texas financier newly involved in animal cloning, to establish a stable of high-level horses. A deal was quickly struck. The moment had come to use the biopsy of the ear that had been taken from Aiken Cura while he was alive, and which since then had been kept cryogenized in the hope of seeing the day when science could perform miracles. The time had come. They would attempt to awaken the animal's cells to bring a genetic copy—a clone—to life.

Cloning: A User's Guide

When we look at the practical side of cloning, the operation loses a bit of the emotion that motivates it. Cloning is a very technical process. One must first isolate unfertilized egg cells called oocytes, which are taken from ovaries removed from mares that have just ended their days at the slaughterhouse. Most often, they have no kinship ties with the animal to be cloned. What follows happens under precision microscopes, using a syringe much thinner than a hair that enables the membrane of each oocyte to be pierced and penetrated to aspirate their nucleus. Emptied of their nuclei—enucleated is the technical term—the oocytes no longer contain any of their chromosomes, or only the chromosome carried by their mitochondria, those mini-organs that enable cells to turn oxygen into energy. But they contain all the ingredients that enable them to remain alive and to undertake, if they are fertilized, a process of development that could in a few days lead to the formation of an embryo.

I followed this stage in the Kheiron Biotech labs in 2018. It requires great concentration and darkness—not to keep the secrets of the process from potential spies, but because light seems to reduce the enucleated oocyte's chances of survival. It happens that these microscopic manipulations also aim to render the following operations easier by removing the thick layer that surrounds the oocytes and protects them, which scientists refer to as the pellucid zone. The oocyte is not only enucleated but also stripped of its natural protection. It is then just a cellular receptacle, ready to undergo the process of embryonic development. It is then put into contact with a cell from the animal to be cloned. That cell could be isolated from a biopsy—skin from the ear, in Aiken Cura's case—and then multiplied in the laboratory to build a stockpile in the event further cloning of the same animal is desired in the future. Biologists call this cell culture.

At this stage, all the ingredients are combined for the remaining operations, which once again will be controlled under a precision microscope at each stage. The cell from the individual to be cloned—the donor cell, which possesses in its nucleus each of the 32 pairs of chromosomes of the future clone—is brought into contact with the enucleated oocyte, also known as the receptor cell. Under the effect of a brief electrical shock, donor and receptor cells then fuse to form only one, which will have inherited chromosomes from the nucleus of the animal to be cloned, as well as all the maternal factors, those transmitted by the oocyte that are necessary for development. The fused cell will then be stabilized in its own culturing medium, and at the right moment will be activated by a chemical cocktail aiming to provide the nucleus with all its potential. It will soon be able to initiate cellular divisions; from the initially single fused mother cell will be issued two daughter cells, each of which will in turn give two daughter cells, and so on. Exponential by nature, this process in principle ends up engendering an embryo containing hundreds of cells after a few days. Some embryos don't have the anticipated shape for normal embryos of the same age, so they are discarded. Those that are in conformity are found under the microscope and taken out of the culture medium to be transplanted into the uterus of surrogate mares.

The technique for transferring embryos in horses has been well mastered, and even if it remains prohibited for certain breeds such as the English Thoroughbred, it is regularly practiced by the equid industry to enable a larger number of mares to bring potential champions into the world every season. In the case of cloning, generally several embryos issued from the same donor are implanted, because not all survive. If an ultrasound reveals that several are developing, only the one that a veterinarian deems to have the best chance of survival will be kept so as not to risk complications during the pregnancy, and the loss of the clone or of its surrogate mother.

Eleven months later, if all goes well, the clone will finally be born—most often in very exclusive, specialized clinics where both the mother and her precious baby are followed using monitoring and ultrasound instruments as sophisticated as those used on humans. Though the technology of cloning has made enormous progress since the birth of Dolly the sheep, who was cloned in 1996, it remains that on average it takes six or seven attempts to obtain a single viable clone—which explains the exorbitant price of the operation, between $50,000 and $100,000 US dollars.

But it is still necessary to wait before the owner can take the coveted clone home, especially if it is destined for a career in polo. It takes a newborn around two years to grow alongside their surrogate mother, who will not lose sight of them until their departure. It is above all the mothers who will nurse them for several months, support them, and provide the comfort necessary for their development. There are also veterinarians and a complete team of professionals who will provide them the best care possible when they leave the nursery and join the stalls of a top-level stable, where the space, equipment, and know-how are abundant. One such top-level stable is Doña Sofía, located around two hours by car from Buenos Aires. There the clones of Kheiron Biotech are pampered, then gradually habituated to the presence of humans as well as that of other horses—and to their sometimes sudden movements, such as those of riders who wield their mallets briskly when they go to strike a polo ball, or when they charge at full speed to make contact before turning on a dime to race off in the opposite direction. It is at the end of this long process that a clone will finally be ready to lead its life as a horse, far from laboratories and closer to the game fields.

The Performance and Health of Clones

Cambiaso wasn't disappointed with his horse's progeny, because Eo1 Cura, the first clone of Aiken Cura, inaugurated the beginning of a new era in polo horse breeding in Argentina—especially after he proved himself by winning the Argentine Open with a complete team of clones. Those clone team members didn't involve copies of Cambiaso's beloved stallion, but those of his beloved mare, Cuartetera. The marketing campaign was unbeatable, and the Crestview Genetics company, which he established with Alan Meeker to clone his champions, was destined to have a spectacular future. It ultimately convinced even the most reluctant who feared clones would be of fragile health, or that they wouldn't perform as well as their original version. After all, hadn't the press reported that the sheep Dolly seemed to show signs of precocious aging?

We know today that Dolly didn't die of a chronic illness, but of a simple infection that worsened, and in the meantime the years that have passed have given us the necessary hindsight to realize that clones on average live as long as their noncloned counterparts. Their physical resemblance is no less astounding, even if when playing the game of "find the seven differences" one might realize that the white spots on the coats of Cuartetera Bo1, Bo2, Bo3, Bo4, Bo5, and Bo6 are not exactly the same size or in the same place (though the pigmentation of skin cells is determined genetically, their migration in embryonic tissue during development is random). According to Cambiaso, the same is true of their abilities in the sport, which are astonishingly reproducible and are capable of transforming the animal into a true machine; even if there seem to be a few subtle differences, only the most seasoned riders might detect them.

A Genetic Lottery

The primary motivation of breeders who use cloning is not necessarily the quest for guaranteed performance, but the opportunity to combine the genes of their champion with those of other champions. Customarily, a mare of Cuartetera's quality would begin her professional career on the playing fields around the age of five, leaving them between five and ten years later once her sporting career was over. She would then be destined for breeding and could bring into the world only a single foal per year, for maybe a few years. Thanks to the technique of transferring an embryo into surrogate mothers, she could technically begin her reproductive duties earlier and become the biological mother of a few additional horses. And because Cuartetera would no longer be a single genetic entity and would be available in a large number of copies of clones, her progeny could quickly reach astronomical numbers. This would enable breeders to try their luck without restriction in the genetic lottery by selecting the clones of Cuartetera and Aiken Cura for breeding new generations blending their exceptional genetic patrimonies, in the hope of giving birth to even more spectacular champions.

They could also do this with other promising stallions, whether clones or not, and multiply their chances of winning the genetic grand prize. They could even clone champions that had been castrated before they were able to reveal their full athletic potential, in order to restore whole animals and enable a lineage that had been destined to disappear. They could even enable horses to continue to reproduce, via their clones, well after their death. Today E01 Cura lives his life in the stable of Los Pingos del Taita de Rio Cuarta, in the Argentine province of Córdoba, far from the polo fields, and his semen is collected and frozen to be used for the artificial insemination of mares throughout the world. The stable refers to itself as "a true cradle for champions." In these conditions, cloning would no longer mean precipitating the end of a sport in which all animals would have become identical; it would prove instead to be a powerfully effective tool enabling the emergence of horses of a new type. A renaissance, if you will.

The argument is seductive, especially since it might find an application outside the admittedly rather select and limited field of polo fields or international competitions of dressage, among which the practice of cloning is also not taboo. I discussed a theoretical idea when I visited the Kheiron Biotech team at the end of 2018: if they were capable of cloning their choice horses in an almost industrial way—their list of orders is not getting shorter, and appears to fill up two years in advance—why couldn't they clone horses whose populations are threatened with extinction? They could begin with Przewalski's horse, native to the Mongolian steppe, which was declared extinct in the wild in 1969 and owes it survival to the concerted efforts of an international program of reproduction in captivity. What better standard-bearer might one dream of for a cause whose importance no one would contest: the preservation of biodiversity?

A Renaissance Beyond the Sports Industry

The idea wouldn't remain in the realm of fiction for long, because on September 4, 2020, a press release announced that a certain Przewalski's foal named Kurt had been born two months earlier. Its name wasn't chosen at random; it was an homage to Kurt Bernischke, not only the tireless protector of Przewalski's horses but the founder of Frozen Zoo, where the tissue of many threatened animal species is kept in liquid nitrogen. Reigning on a shelf in my office is one of his statuettes in the effigy of the horse he loved so much, which his widow gave to me when I was at a conference in San Diego. Kurt is the clone of a Przewalski's horse named Kuporovic. It is not, however, at Doña Sofía or at Los Pingos del Taita de Rio Cuarta where one might marvel at seeing it stretch its legs and frolic, as in the video clip accompanying the press release, but at the veterinary clinic of Timber Creek Veterinary Hospital in Texas, 500 kilometers east of Albuquerque. And for a good reason: the clone wasn't born in the labs of Kheiron Biotech or Crestview Genetics but in those of ViaGen Pets, a Texan company specializing in the cloning of horses and pets—a market that is apparently in full expansion in North America.

Kuporovic I—the original Kuporovic, so to speak—was born captive in 1975 in England, and three years later was transferred to the San Diego Zoo, one of the world temples of conservation biology. He lived there until 1998. At his birth, Kurt was forty-five years older than his original version. The cells that were used for the cloning came straight from the Frozen Zoo. I was fortunate to be able to visit that biobank of cryogenized tissue with its director, Oliver Ryder, at the beginning of 2014 while we were working together to characterize the genetic diversity of Przewalski's horses.

Kuporovic's cellular lineage derives from a biopsy that had been carried out in 1980; its scientific name is SB615. I know it well because several years earlier Ryder had already sent me DNA extracts from it so that my lab could sequence its genome. To complete our analysis, Ryder had also taken care to add those of some fifteen other lineages meticulously selected to represent most of the great genetic families known to have particularly structured the genealogy of Przewalski's horses in captivity throughout the twentieth century. For the first time, the nature of our sampling gave us the possibility of characterizing the genetic pool of that population of endangered horses and the extent of its diversity on the scale of its genome in its entirety. Furthermore, to perfect our study, we had worked to enhance the extracts of museum specimens prepared between the end of the nineteenth century and the beginning of the twentieth, in order to present the state of genetic loci of the contemporary situation compared to what it had been before the decline that had precipitated the disappearance of that population in the wild.

From Cloning to Time Travel

Some of our museum specimens were particularly important. For example, we sampled that of an animal that died in 1878 and caused Western science to discover *Equus przewalski*; it is considered to be the stallion of reference that defines the species in its entirety (this is called a holotype)—the most Przewalski of all the horses, in other words. Beyond the holotype and other key pieces from the collections of the Zoological Museum in St. Petersburg, Russia, our sampling also contained specimens from the Julius Kühn collections of the Museum of Livestock Science in Halle (Saale) in Germany, which at the beginning of the twentieth century was one of the first European points of entry for Przewalski's horses captured in Mongolia. The place was home to animals whose descendance proved particularly prolific in the history of the reproduction of an animal in captivity; their influence on the genealogy of Przewalski's horses was significant, including some like Theodore, who was born in 1905, in all likelihood from a domesticated Mongolian mare. If that were proven true, it would be a hybrid, whose father alone would be a true Przewalski's horse. The genealogy of the Przewalskis would then contain domesticated blood.

Our analysis went in that direction. With close to a quarter of its genome of domesticated origin and three-quarters issued from a Przewalski, it could not mathematically be the son of a domesticated mother and a Przewalski father. For that it would have been necessary to find 50 percent of each. But with such figures, there was no doubt that one of its grandparents was of domesticated extraction. Theodore wasn't a direct hybrid, but the son of a hybrid; domesticated blood flowed in his veins. By extension, that meant there must be many current Przewalski's horses in the same situation, for a very simple reason: Theodore appeared at the top of the genealogy of horses reproduced in captivity, at a generation or two from the oldest founding fathers. His children, as well as the children of his children—all of his descendants—contributed to transmitting genes of domesticated origin into the current population of these threatened horses. It wasn't globally pure, but was in part mixed.

Back to the Future

Among the many descendants of Theodore are Kuporovic, and consequently Kurt, who is nothing other than his true genetic copy. Like many of his counterparts, the young clone is therefore not a pure-blood Przewalski. According to our estimates, his genome would contain 11 percent to 28 percent domesticated ascendance. Does this mean we brought Kurt into the world for nothing? Certainly not! His birth had above all proved that cloning technology was mature enough to enhance the arsenal of conservation biology and cause the rebirth more than forty years apart of two genetically identical individuals. The history of crossings in captivity has shown that some of Kuporovic's descendants kept as much original domestic genetic ancestry as he did—if their parents contained equivalent proportions of it. This same history has also resulted in diluting this proportion in others among his descendants each time one of the parents had less domesticated blood than Kuporovic. Ultimately, the history of the crossings even led to an increase of that proportion in some animals—this would be the case if Kuporovic had reproduced with a Przewalski's horse closer to Theodore in the genealogy, because there would have then appeared in his genome a larger portion of DNA of domesticated origin. All of this is to say that with Kurt we have the possibility of in part reshaping the history of the crossings as if they had not occurred, as if to correct and prevent certain crossings—for example, those augmenting the portion of domesticated DNA, or those of particularly consanguine nature.

Our work also showed that consanguinity was generally greater in Przewalski's horses today than it was at the end of last century, and that it could sometimes reach very disturbing proportions in some of them. If I say disturbing, it is because consanguinity by definition reveals the presence of crossings between related individuals. The same especially harmful mutation that might exist in one copy in both of the two parents without causing pathologies in them (this is called a recessive harmful mutation) could well be found in two copies in one of their children, if transmitted by each of the two parents. The foal with no other genetic copy than the harmful mutation could be fully subjected to its harmful effects, and present heightened risks of developing chronic illnesses or certain forms of sterility. When populations are very large, such mutations in general remain relatively rare, even very rare; the chance that two healthy carriers reproduce together is even more unlikely. But with populations of very limited size, as among Przewalski's horses, the chance that such mutations are maintained, or even become more frequent over generations, is far from small. The risk that they cross and that a newborn develops a pathology is increased.

To be convinced of this, let's look at an extreme example. Let's imagine a group of ten Przewalski's horses made up of five males and five females. In this group, only one male and one female are healthy carriers of a harmful mutation. The frequency of that mutation within the group would be of two copies out of twenty possible copies, or 10 percent (each individual carries two copies of the same gene, one from its mother, the other from its father). In the game of reproduction only carrying individuals reproduce (the size of the population of reproducing individuals is limited here to the extreme, two individuals) and let's say that their unions, in their lifetime, lead to the birth of eight foals. On average, a quarter of them will carry two copies of the harmful mutation and one quarter won't carry it, while the remaining half will carry only a single copy. A quick calculation is enough to realize that the frequency of the harmful mutation can leap from 10 percent to 50 percent in a single generation. While neither of the parents suffered from its harmful effects (they both had a nonharmful copy in their genome), two of the eight animals issued from their union will likely suffer its consequences. We can easily see how the randomness of genealogies, especially when the populations are small, can compromise the health of the descendancy.

That isn't all. Our work has also shown that the proportion of harmful mutations is higher in Przewalski's horses—so the risk, as theoretical as it might have appeared at first, had every chance of having regrettable practical consequences. With Kurt, another family tree than that of Kuporovic could be created, with new relatives and descendants, multiplying the chances to reassort its genes in many different ways. Replaying the same game of reshuffling the genealogy and genes with other clones of Przewalski's horses would increase the combinations even further, which could provide conservation biologists with a new key asset. Another asset would derive from the cloning of mares (instead of stallions), because they would ultimately enhance the reproduction potential of the threatened population: more foals and fillies would be born every year than what is currently possible with the limited number of existing mares.

Finally, if our work has shown that a predominant portion of the captive genealogy of Przewalski's horses carried in it a genetic legacy of domesticated origin, it has also revealed that there remained a portion that did not. Therefore, the horses concerned resembled those that, like the holotype, lived at the end of the nineteenth century. The good news is that Oliver Ryder's Frozen Zoo contains cellular lineages of some of these horses. We could then use them to bring genetic clones to life to support Kurt and enhance the reproduction plans for horses in captivity. This is an appeal to repeat either the Kurt or Kuporovic experiment, depending on whether one prefers the true copy or the original version. It would however be important not to exclusively favor the biologically purest Przewalski's horses, because then one would certainly and dramatically reduce available genetic diversity, which is already limited. In doing so, one would also risk increasing the level of consanguinity in the population, because the purest individuals go back to a small number of identical founders and are therefore related at diverse but not negligible levels. This could be all the more disastrous, as the proportion of harmful mutations present in their genome is slightly higher than that which one finds in Kuporovic and his cousins. It seems that the future is not in irrational cloning full-speed ahead, but in a concerted action that initially involves the sequencing of the genome of candidates for cloning as well as the evaluation of the most promising crossings. Technology would therefore be in the service of a rational evaluation of the best options offered to conservation biologists.

Copies: No Longer True, but Edited

But might we imagine going one step further in the use of the tools of genomics? What if we no longer confined ourselves simply to copying the genetic information contained in the mobilized cells via cloning, but begin to edit that information beforehand? We could touch it up, in other words, in places of the genome known to control certain characteristics, by augmenting it with attributes the original didn't have. Here we would be realizing the dream of most breeders, since the characteristics aimed for would not be recombined at random and selected following a long process of crossings stretching over many generations, but would be immediately acquired in a single stage. It's the ultimate promise: you can allow the breeding to reach a level of high precision without having to wait. As it turns out, this experiment isn't all that far from the realm of science fiction, because the step was already essentially taken in a lab in 2020. Again, it was for polo horses.

The method used was that of molecular scissors—a discovery that, the same year, earned Jennifer Doudna (of the US) and the Emmanuelle Charpentier (from France) the Nobel Prize in chemistry. The goal of the experiment carried out on a horse was to verify if the tools we possess currently are capable of replacing, in the donor cells of a future clone, a very short sequence localized in the *MSTN* gene with a sequence carried by another horse. The choice of the associated sequence was not random, since it targeted a gene coding for the protein myostatin, whose role affects muscle growth and development, as we've seen in chapter 11. Remember: individuals who naturally carry the chosen sequence in two copies on average have a greater muscle mass and obtain better performances in sprint races. The 2020 study established the experimental conditions that enable the introduction of the chosen sequence precisely at the desired place, and nowhere else. This surgical operation on the genome doesn't present the risk of affecting functions other than the one targeted; nothing prevents these edited cells from being used in future cloning and in leading to the birth of an edited clone. This might result in a super-athlete with enhanced musculature not through the use of steroids, but from the genes of another super-athlete. Someday, it might potentially produce a horse with some of the harmful mutations it carries and might transmit edited out, such as those that affect the *PPIB* gene or the *GBE1* gene, which can lead to very severe conditions of the skin or the metabolism in domesticated horses and many other conditions as well.

Given all that, why not imagine that we could eliminate the most harmful mutations in Kuporovic before cloning him, and repeat the operation in cloning candidates to alleviate the genetic burden that plagues Przewalski's horses? We could also envision reintroducing into their genome some of the non-harmful mutations we have been able to characterize among the bone fragments of the Botai horses, to boost the currently limited genetic diversity of these endangered horses by adding the diversity that characterized their direct ancestors. Given that our work has shown a significant decline in genetic lineages since the end of the nineteenth century, the experiment of 2020 carried the seeds of a program with dizzying ramifications.

An Uncertain Future

It was with all these questions in my head that I met Daniel
Sammartino, founder of Kheiron Biotech. He invited me to his
splendid home set in the heart of a golf club located not so far
from the laboratory where cloned and edited embryos are prepared, for the purpose of a scientific documentary about animal
cloning. I couldn't resist asking those questions. Kheiron's policy is that there is a red line that must never be crossed: that of
genetically modified organisms, more commonly known by the
very polemical acronym GMO. To avoid this, one needs to
agree never to provide a genetic edition other than versions of
the genetic information that have already been found in a horse.
The clone in its edited version could then not be considered a
GMO, since 100 percent of the genetic information that it carries would come from real horses. No fragment would cross the
natural barriers that separate species and living beings, unlike
GMOs.

Though the position is defended on technical and ethical
levels, one cannot help noticing that it has also been revealed
to be useful on the judicial and commercial levels, because the
legislation that surrounds the fabrication and commercializa-
tion of GMOs is particularly strict. By remaining radically distant from the GMO label, Sammartino, as a good CEO, orients
the future of his biotech company toward a much more immediately maneuverable market. Will the other biotech companies
competing in the cloning and genetic editing market make the
same choice, and if so, for how long? Will horses edited to be
genetically augmented become the norm in the sport? Will
other sports be seduced? Will the practice be limited to the
domains of sport, or will it be propagated with the motive of
preserving a declining biodiversity? The future will tell us, but

one thing is certain: the responses that will be given to all these questions will depend in part on what the horses of the future will look like.

This book has shown how the most powerful tools in genomics enable us to retrace the multimillennial history of domesticated horses. And what is more, rather than being just decipherers of genetic traces extracted from the depths of the past, those same tools today are capable of writing the history of our horses in the future.

13

Epilogue

The Horse and Me

My personal history with horses didn't really begin until the spring of 2010. I had left France to set up my research team at the University of Copenhagen. Three years later, we sequenced the genome of an animal close to 700,000 years old. It was a horse, but to be honest it was the age of the animal, rather than its species, that had launched us on this adventure. It was a time for exploring what the so-called new generation of sequencing technologies could truly do: How far could we take them? However, that challenge took so much energy and mobilized so many resources that I almost mechanically ended up being able to fix my attention solely on that species and its evolution. We were at a time when science knew only one horse genome, that of Twilight, an English Thoroughbred mare born on the Cornell University campus in Ithaca, New York. In a single study we had just characterized almost ten genomes, and one of them exceeded by more than a half-million years a time barrier that at the time we thought was uncrossable. Now here we are fifteen years later, back in France, but just as motivated as on the first day. But we are surely on our way for a new decade of tracking the genomes of horses from all ages, so inexhaustible does what remains to be studied appear.

However, my personal experience with horses could have begun much earlier. In middle school, scheduling played a role; I was fortunate to attend a school where one afternoon a week was set aside to practice a sport. For four years, I could spend Tuesday afternoon either horseback riding or sailing. I didn't hesitate: between tacking from the Chateau d'If to the Iles du Frioul or cleaning up the dung of an animal that had not yet become my traveling companion, I chose sailing. The horse had to continue to do what it does best: wait patiently for the day when our paths would cross again. Some dozen years later, in graduate school, the opportunity presented itself again; this time I happened to meet Paul Sondaar and Véra Eisenmann, two world-renowned paleontologists, both specialists in equid evolution. Sondaar did attempt to get my foot into a stirrup by telling me about the history of the horse family in the Americas, of which I was essentially ignorant; Eisenmann also tried to influence me, as someone who had devoted her professional life to the equids of world, past and present. But at that time, ge-nomics, and in particular that applied to archeological remains, was not where it is today. We could sequence a few gene frag-ments: always the same ones, with which we had to be content to estimate genetic affinities between species and populations with, let's admit it, quite relative precision. I wasn't ready to take the leap and devote most of my time to just that species. I de-voted some time to the horse occasionally, in between work on the Neanderthals, cave bears, or very distant cousins of horses, odd-toed ungulates such as woolly rhinoceroses. It took me al-most another decade to realize that the horse represented a formidable subject for study. The animal had to wait for our third encounter. The opportunity presented itself in the remains of Thistle Creek, trapped in the Canadian permafrost for more than a half-million years.

A Fascinating Object of Study

My interest in horses was initially theoretical and academic; it was primarily because it was an object of study that still remained mysteriously unknown at the beginning of the twenty-first century. The more I learned about the subject, the more I realized the extent of the questions that remained to be answered. The horse family had occupied an important place in the science of evolution since the work of George Gaylor Simpson in the 1950s, revealing over 60 million years of anatomical changes and providing us with one of the fundamental and indisputable proofs that evolution is not a theory, but a reality. And yet we didn't have the answers to such simple questions as how many species still existed 25,000 years ago. Furthermore, the horse may have represented a major tipping point in the history of civilizations, but specialists were still struggling to find traces of the cradle of its domestication. Whereas entire books had been devoted to describing riding techniques and horse breeds, we basically knew nothing of their history and their true origin. It didn't make sense. All of that was dizzying, because paradoxically, it seemed that everything, or almost everything, had been written on the subject without ever providing definitive answers. Where should one begin?

I remember exactly the moment when I began to prepare my candidacy for a European grant—one of those that are awarded by the European Council for Research. The process requires the exact opposite of what financial organizations generally request: a maximum of risk-taking. It involves proposing a project of scientific disruption, something essential but difficult to the point of never having been resolved, something that would have every reason to end in failure, but which, if successful, would have a fundamental impact. When reflecting on what the project entailed, I remember having asked myself the question in those terms and wondering: "Am I really ready to spend the next ten years of my career on this subject?" The questions I anticipated broaching would imply complete commitment. The first version of the project wasn't financed, but I tried again the following year and was funded by the Pegasus Project, which would support us financially until the end of 2022 (now extended to 2029, with the Horsepower project). We had decided to include not only paleogeneticists capable of sequencing and analyzing genomes but also an entire team of archeologists, prehistorians, historians, and linguists. If science in its disciplinary cubicles had not yet been able to pierce the mysteries of the deep history of this animal, there was no reason that we would be able to break through the dams if we didn't combine expertise and approaches. This book will enable everyone to judge if the challenge has been met.

The Pegasus Project provided me the opportunity to have a fourth encounter with the horse, one that was, perhaps, when I think about it, the most important of all. To be honest, the opportunity was inspired more by my wife than by the project itself, because she was the one who had the idea of signing me up for riding lessons at the riding school, coincidentally located very close to where we lived. At the age of forty-two. In her opinion, I could not continue to think horse, talk horse, and eat horse if I had never actually been on one. It defied all reason. How could I think that reading, theoretical reflection, and abstraction could be enough when it involved understanding the profound reasons that linked our humanity to a nonhuman living being? Even if my wife wouldn't get closer than 10 meters to the animal for anything in the world—she is frightened of it—it was obvious to her that I needed to have experience with the animal myself, learn how one approaches it, how it behaves, how it listens, and to experience the sensation of being on its back, to hear it breathe and feel it running fast.

The Discovery of an Animal to Which I Feel Extremely Close

How right she was! For a long time, I was proud not to feel an emotional attachment to that animal; I even considered that to be an advantage. Times have changed, but if I put myself in the skin of the person I was not so long ago, in the beginning my sympathies went more easily toward elephants or crocodiles. It wasn't uncommon that I would begin my scientific presentations at international meetings by explaining why I was passionate about the history of horses, even though I wasn't a rider. It always surprised my colleagues and my friends that I could practice the profession I had chosen without having grown up in stables and spending my adolescence galloping around, and without having become a racetrack enthusiast. The example of Blake Shinn, who lost his trousers at full speed on the Australian track during the Hyland Race Colours Plate and had to finish the race with a bare behind, was not of the kind to make me change my mind. I justified myself; deep down, I sincerely thought that not being a rider was precisely what enabled me, perhaps more than others, to rid myself of any *a prioris*, to be more objective, and not to reject out of hand those hypotheses that came from off the beaten path. Whether the Arabian horse or the American horse had a certain history or a different one for me changed nothing. As far as I was concerned it was a subject of study, not a personal interest. I could envision everything without restraint, uninhibited.

If I haven't abandoned that point of view—I remain a scientist guided by experimental data and only by that—today I must admit that interacting with the real animal, in flesh and bone, has enabled me to go even further in my research. If the animal is an object of study, it is not just an object. A strictly clinical approach would only miss an important part of what had governed the choices of breeders of the past—because there are in fact movements and ways of doing things that the animal will never accept, and there are others that, on the contrary, more easily bring us closer to it. Understanding how different breeding techniques and different schools of thought during its history have been able to contribute to shaping the animal beyond its mere immediate utility has become a completely different subject, now at the heart of the interests of my lab.

The Horse as a Bridge

While taking me to the four corners of the planet, the Pegasus Project has taught me something that will forever remain essential: the degree to which we are all, from Argentina to China, including Mongolia, central Asia, and the vast expanse of Russia, perhaps even in spite of ourselves, filled with nostalgia when we talk about this animal. It is moreover perhaps what pushes some, like Bob Long, to set off at the age of seventy on races as mad as the Mongol Derby, the longest in the world, with its 1,000 kilometers through the Mongolian steppe, dotted with stages inspired by the Genghis Khan postal relay; to have the feeling of reliving the sensations of eras that are today long gone, but not really so distant. In other words, to resuscitate a bit of the scent of the great horse races in history. I hope this book, at its modest level, has contributed to that. In essence, Peter Goullart, the Russian explorer of the 1940s, said exactly the same thing when he found himself at the doors of the Ancient Tea and Horse Road. To quote just one passage in his *Forgotten Kingdom*, he was essentially telling us: "If it happened that a nuclear cataclysm destroyed all our modern means of communication and transportation, we can always count on this discreet ally, our oldest friend, the horse, to once again forge connections between those peoples and nations that have been temporarily torn apart."

The Pegasus Project has given me the opportunity to verify that a million times. Wherever I go, with or without an interpreter, I have always been able to note the same thing: the horse remains what it has been for almost our entire history, a vector for bringing people together. I remember the smile the animal provoked in the breeders of the Khomiin Taal Plateau in Mongolia. I remember Lolo, gaucho working at Doña Sofía, and how proud he was when he showed me how he took care of his foals so they could soon make their entrance onto the polo fields. I remember Pavel Kuznetsov when he pointed to the world of tarpans. And I remember Alexey Tishkin's excitement when he showed me the drawings of horses that humans had engraved on the rocks of Altai thousands of years ago. Isn't that also what the Lakota Oglala, an Indigenous Sioux tribe in the Americas, had revealed to me in making *šuŋkawakaŋ*, the horse, a nation—their equal? I remember what Yvette Running Horse Collin told me during our very first encounter, even before she joined my lab: "Follow the horse, Ludovic," because it shows us the way. May it continue to do so for centuries to come.

BIBLIOGRAPHY

Chapter 1. Prologue

Association TAKH. "Safeguarding Przewalski's horses." Accessed January 12, 2025. https://www.takh.org/.

Baratay, E. *Bêtes des tranchées. Des vécus oubliés.* CNRS Éditions, 2013, p. 256.

Bold, B. *Eques mongolica. Introduction to Mongolian Horsemanship.* Bold & Boldi, 2012, p. 288.

Buffon, G. L. L. *Histoire naturelle de Buffon.* Place des Victoires, 2022, p. 448.

Di Marco, L. A. *War Horse: A History of the Military Horse and Rider.* Westholme Publishing, 2012, p. 416.

Freeberg, E. "How a flu virus shut down the US economy in 1872 by infecting horses." *The Conversation,* December 3, 2020. https://theconversation.com/how-a-flu-virus-shut-down-the-us-economy-in-1872-by-infecting-horses-150052.

Kelekna, P. *The Horse in Human History.* Cambridge University Press, 2009, p. 460.

Khadka, R. *Horse Population, Breeds and Risk Status in the World: A Study Based on Food and Agriculture Organization Database Systems: FAOSTAT and DAD-IS.* Lambert Academic Publishing, 2011, p. 80.

Koselleck, R. "Der Aufbruch in die Moderne oder das Ende des Pferdezeitalters." In *Historikerpreis der Stadt Münster 2003.* Münster City Historian Prize booklet, 2003, pp. 23–37.

Librado, P., N. Khan, A. Fages, et al. "The origins and spread of domestic horses from the Western Eurasian steppes." *Nature* 598 (2021), pp. 634–640.

McShane, C. *The Horse in the City: Living Machines in the Nineteenth Century.* Johns Hopkins University Press, 2007, p. 274.

Milhaud, C. *1914–1918. L'autre hécatombe,* Belin, Équitation, 2017, p. 302.

Orlando, L. *L'ADN fossile. Une machine à remonter le temps.* Odile Jacob, 2020, p. 252.

Orlando, L., R. Allaby, P. Skoglund, et al. "Ancient DNA analysis." *Nature Reviews Methods Primers* 1 (2021), pp. 1–26.

Raulff, U. *Farewell to the Horse. The Final Century of Our Relationship*. Penguin Books, 2018, p. 480.

Sidnell, P. *Warhorse: Cavalry in Ancient Warfare*. Bloomsbury Publishing, 2007, p. 376.

Chapter 2. The Origin Horse

Berger, J. "Organizational systems and dominance in feral horses in the Grand Canyon." *Behavioral Ecology and Sociobiology* 2 (1977), pp. 131–146.

Chechushkov, I. V., and P. A. Kosintsev. "The Botai horse practices represent the neolithization process in the Central Eurasian steppes: Important findings from a new study on ancient horse DNA." *Journal of Archaeological Science: Reports* 32 (2020), 102426. https://doi.org/10.1016/j.jasrep.2020.102426.

Der Sarkissian, C., L. Ermini, M. Schubert, et al. "Evolutionary genomics and conservation of the endangered Przewalski's horse." *Current Biology* 25 (2015), pp. 2577–2583.

Dudd, S. N., and R. P. Evershed. "Direct demonstration of milk as an element of archaeological economies." *Science* 282 (1998), pp. 1478–1481.

Fages, A., A. Seguin-Orlando, M. Germonpre, and L. Orlando. "Horse males became over-represented in archaeological assemblages during the Bronze Age." *Journal of Archaeological Science: Reports* 31 (2020), 102364. https://doi.org/10.1016/j.jasrep.2020.102364.

Gaunitz, C., A. Fages, K. Hanghøj, et al. "Ancient genomes revisit the ancestry of domestic and Przewalski's horses." *Science* 360 (2018), pp. 111–114.

Kempson, I. M., and D. A. Henry. "Determination of arsenic poisoning and metabolism in hair by synchrotron radiation: The case of Phar Lap." *Angewandte International Edition Chemie* 49 (2010), pp. 4237–4240.

Krueger, K., B. Flauger, K. Farmer, and C. Hemelrijk. "Movement initiation in groups of feral horses." *Behavioural Processes* 103 (2014), pp. 91–101.

Levine, M. "Botai and the origins of horse domestication." *Journal of Anthropological Archaeology* 18 (1999), pp. 29–78.

Linklater, W. L. "Adaptive explanation in socio-ecology: Lessons from the *Equidae*." *Biological Reviews* (75), 2000, pp. 1–20.

Lowry, B. *Killing Phar Lap: An Untold Part of the Story*. AuthorHouse, 2014, p. 116.

Moehlman, P. D. *Equids: Zebras, Asses, and Horses. Status Survey and Conservation Action Plan*. IUCN, Gland, 2002, p. 189. https://portals.iucn.org/library/sites/library/files/documents/2002-043.pdf.

Negara, K. "Killing Phar Lap: A forensic investigation." Apple podcast, 2020–2022. https://podcasts.apple.com/us/podcast/killing-phar-lap-a-forensic-investigation/id1536743152.

Olsen, S., B. Bradley, D. Maki, and A. Outram. "Community organization among Copper Age sedentary horse pastoralists of Kazakhstan." In *Proceedings of the 2002 University of Chicago Conference on Eurasian Archaeology*, edited by D. L. Peterson, L. M. Popova, and A. T. Smith. Brill Academic Publishers, 2006, pp. 89–111.

Olsen, S. L. "Early horse domestication on the Eurasian steppe." In *Documenting Domestication: New Genetic and Archaeological Paradigms*, edited by S. L. Olsen and M. A. Zeder. University of California Press, 2006, pp. 245–269.

Orlando, L., A. Ginolhac, G. Zhang, et al. "Recalibrating *Equus* evolution using the genome sequence of an early Middle Pleistocene horse." *Nature* 499 (2013), pp. 74–78.

Outram, A. K., R. Bendrey, R. P. Evershed, L. Orlando, and V. F. Zaibert. *Rebuttal of Taylor and Barrón-Ortiz 2021 "Rethinking the evidence for early horse domestication at Botai."* Zenodo, 2021. https://ore.exeter.ac.uk/repository/bitstream/handle/10871/126599/Outram%20et%20al%20Rebuttal%20of%20Taylor.pdf.

Outram, A. K., N. A. Stear, R. Bendrey, et al. "The earliest horse harnessing and milking." *Science* 323 (2009), pp. 1332–1335.

Scanlan, L. *The Horse God Built: The Untold Story of Secretariat, the World's Greatest Racehorse*. St Martin's Griffin, 2008, p. 335.

Taylor, W. T. T., and C. I. Barrón-Ortiz. "Rethinking the evidence for early horse domestication at Botai." *Scientific Reports* 11 (2021), p. 7440. https://doi.org/10.1038/s41598-021-86832-9.

Zhang, X., M. Hirschfeld, R. Schafberg, H. Swalve, and B. Brenig. "Skin exhibits of Dark Ronald XX are homozygous wild type at the Warmblood fragile foal syndrome causative missense variant position in lysyl hydroxylase gene PLOD1." *Animal Genetics* 51 (2020), pp. 838–840.

Chapter 3. The Other Origin of the Horse

Anthony, D. W. *The Horse, the Wheel, and Language. How Bronze-Age Riders from the Eurasian Steppes Shaped the Modern World*, Princeton University Press, 2007, p. 568.

Anthony D. W., and D. Brown. "The secondary products revolution, horse-riding, and mounted warfare." *Journal of World Prehistory* 24 (2011), pp. 131–160.

Anthony, D. W., A. A. Khokhlov, S. A. Agapov, et al. "The Eneolithic cemetery at Khvalynsk on the Volga River." *Praehistorische Zeitschrift* 97 (2022), p. 46.

Benecke, N. "On the beginning of horse husbandry in the Southern Balkan Peninsula. The horse bones from Kırklareli—Kanligeçit (Turkish Thrace)." In Mashkour M., *Equids in Time and Space. Papers in Honour of Véra Eisenmann*, edited by M. Mashkour. Oxbow Books, 2006, pp. 13–24.

Chechushkov, I. A., A. S. Yakimov, O. P. Bachura, Y. Chuen Ng, and E. N. Goncharova. "Social organisation of the Sintashta-Petrovka groups of the Late Bronze Age and a cause for origin of social elites (based on materials of the settlement of Kamenny Ambar)." *Stratum Plus* 2 (2018), pp. 149–166.

Chechushkov, I. V., and A. V. Epimakhov. "Eurasian steppe chariots and social complexity during the Bronze Age." *Journal of World Prehistory* 31 (2018), pp. 435–483.

De Barros Damgaard, P., R. Martiniano, J. Kamm, et al. "The first horse herders and the impact of early Bronze Age steppe expansions into Asia." *Science* 360 (2018), p. eaar7711. https://doi.org/10.1126/science.aar7711

Drews, R. *Militarism and the Indo-Europeanizing of Europe.* Routledge, 2017, p. 294.

Fages, A., A. Seguin-Orlando, M. Germonpré, and L. Orlando. "Horse males became over-represented in archaeological assemblages during the Bronze Age." *Journal of Archaeological Science: Reports* 31 (2020), 102364. https://doi.org/10.1016/j.jasrep.2020.102364.

Hahn, M. *Molecular Population Genetics.* Oxford University Press, 2018, p. 352.

Kitov, E. P., A. A. Khokhlov, and P. Ankusheva. "Paleoanthropological Data as a source of reconstruction of the process of social formation and social stratification (based on the Sintashta and Potapovo sites of the Bronze Age)." *Stratum Plus* 2 (2018), pp. 149–166.

Koryakova, L., and A. Epimakhov. *The Urals and Western Siberia in the Bronze and Iron Ages.* Cambridge University Press, 2007, p. 408.

Kosintsev, P., and P. Kuznetsov. "Comment on "The earliest horse harnessing and milking," *Tyragetia* 7 (2013), pp. 405–408.

Kyselý, R., and L. Peske. "Horse size and domestication: Early equid bones from the Czech Republic in the European context." *Anthropozoologica* 51 (2016), pp. 15–39.

Leonardi, M., F. Boschin, K. Giampoudakis, et al. "Late Quaternary horses in Eurasia in the face of climate and vegetation change." *Science Advances* 4 (2018), p. eaar5589. https://doi.org/10.1126/sciadv.aar558.

Levine, M. A. "Dereivka and the problem of horse domestication." *Antiquity* 64 (2015), pp. 727–740.

Levine, M. A. "The origins of horse husbandry on the Eurasian steppe." In *Late Prehistoric Exploitation of the Eurasian Steppe,* edited by M. A. Levine et al. McDonald Institute for Archaeological Research, 1999, pp. 5–58.

Librado, P., N. Khan, A. Fages, et al. "The origins and spread of domestic horses from the Western Eurasian steppes." *Nature* 598 (2021), pp. 634–640.

Lindner, S. "Chariots in the Eurasian steppe: A Bayesian approach to the emergence of horse-drawn transport in the Early Second Millennium BC." *Antiquity* 94 (2020), pp. 361–380.

Olsen, S. L., and M. Zeder. *Documenting Domestication. New Genetic and Archaeological Paradigms.* University of California Press, 2006, p. 375.

Takahashi, A., and K. A. Miczek. "Neurogenetics of aggressive behavior: Studies in rodents." *Current Topics in Behavioral Neurosciences* 17 (2014), pp. 3–44.

Tikker, L., P. Casarotto, P. Singh, et al. "Inactivation of the GATA cofactor ZFPM1 results in abnormal development of dorsal raphe serotonergic neuron subtypes and increased anxiety-like behavior." *Journal of Neuroscience* 40 (2020), pp. 8669–8682.

Vigne, J. D. "The origins of animal domestication and husbandry: A major change in the history of humanity and the biosphere." *Comptes rendus de l'Académie des sciences biologie* 334 (2011), pp. 171–181.

Warmuth, V., A. Eriksson, M. A. Bower, et al. "European domestic horses originated in two Holocene refugia." *PLoS One* 6 (2011), e18194. https://doi.org/10.1371/journal.pone.0018194.

Wilkin, S., A. R. Ventresca-Miller, R. Fernandes, et al. "Dairying enabled Early Bronze Age Yamnaya steppe expansions." *Nature* 598 (2021), pp. 629–633.

Chapter 4. The Horse of the Apocalypse

Allentoft, M. E., M. Sikora, K.-G. Sjögren, et al. "Population genomics of Bronze Age Eurasia." *Nature* 522 (2015), pp. 167–172.

Anthony, D. W. *The Horse, the Wheel, and Language. How Bronze-Age Riders from the Eurasian Steppes Shaped the Modern World.* Princeton University Press, 2007, p. 568.

Beckes, R. S. P., and M. De Vaan. *Comparative Indo-European Linguistics: An Introduction.* John Benjamins Publishing, 1995, p. 398.

Davis, R. H. C. *The Medieval Warhorse. Origin, Development and Redevelopment.* Thames and Hudson, 1989, p. 144.

Demoule, J.-P. *Mais où sont passés les Indo-Européens? Le mythe d'origine de l'Occident.* Seuil, 2017, p. 848.

DiMarco, L. A. *War Horse: A History of the Military Horse and Rider.* Westholme Publishing, 2012, p. 424.

Drews, R. *Early Riders: The Beginnings of Mounted Warfare in Asia and Europe.* Routledge, 2004, p. 218.

Furholt M. "Massive migrations? The impact of recent aDNA studies on our view of Third Millennium Europe." *European Journal of Archaeology* 21 (2018), pp. 159–191.

Furholt, M. "Mobility and social change: Understanding the European Neolithic Period after the archaeogenetic revolution." *Journal of Archaeological Research* 29 (2021), pp. 481–535.

Gazagnadou, D. "Les étriers. Contribution à l'étude de leur diffusion de l'Asie vers les mondes iranien et arabe." *Techniques et Culture* 37 (2001), pp. 155–171.

Gimbutas, M. *The Prehistory of Eastern Europe. Part I: Mesolithic, Neolithic and Copper Age Cultures in Russia and the Baltic Area.* Peabody Museum, 1956, p. 241.

Goldberg, A., T. Günther, N. A. Rosenberg, and M. Jakobsson. "Ancient X chromosomes reveal contrasting sex bias in Neolithic and Bronze Age Eurasian migrations." *Proceedings of the National Academy of Sciences USA* 114 (2017), pp. 2657–2662.

Goldberg, A., T. Günther, N. A. Rosenberg, and M. Jakobsson. "Reply to Lazaridis and Reich: Robust model-based inference of male-biased admixture during Bronze Age migration from the Pontic-Caspian steppe." *Proceedings of the National Academy of Sciences USA,* 114 (2017), p. e3875–e3877.

Haak, W., I. Lazaridis, N. Patterson, et al. "Massive migration from the steppe was a source for Indo-European languages in Europe." *Nature* 522 (2015), pp. 207–211.

Hyland, A. *The Medieval Warhorse: From Byzantium to the Crusades.* Sutton Publishing, 1996, p. 215.

Librado, P., N. Khan, A. Fages, et al. "The origins and spread of domestic horses from the Western Eurasian steppes." *Nature* 598 (2021), pp. 634–640.

Librado, P., G. Tressières, L. Chauvey, et al. "Widespread horse-based mobility arose around 2200 BCE in Eurasia." *Nature* 631 (2024), pp. 819–825.

Lindner, S. "Chariots in the Eurasian steppe: A Bayesian approach to the emergence of horse-drawn transport in the Early Second Millennium BC." *Antiquity* 94 (2020), pp. 361–380.

Littauer, M. A., and J. H. Crouwel. "The origin of the true chariot." *Antiquity* 70 (1996), pp. 934–939.

MacHugh, D., G. Larson, and L. Orlando. "Taming the past: Ancient DNA and the study of animal domestication." *Annual Reviews of Animal Biosciences* 5 (2017), pp. 329–351.

Novembre, J. "Ancient DNA steps into the language debate," *Nature,* 522, 2015, pp. 164–165.

Papac, L., M. Ernée, M. Dobeš, et al. "Dynamic changes in genomic and social structures in Third Millennium BCE Central Europe." *Science Advances* 7 (2021), eabi6941. https://doi.org/10.1126/sciadv.abi6941.

Pellard, T., L. Sagart, and G. Jacques. "L'indo-européen n'est pas un mythe." *Bulletin de la Société de linguistique de Paris* 113 (2018), pp. 79–102.

Piazza, A., S. Rendine, E. Minch, P. Menozzi, J. Mountain, and L. L. Cavalli-Sforza. "Genetics and the origin of European languages." *Proceedings of the National Academy of Sciences USA* 92 (1995), pp. 5836–5840.

Racimo, F., J. Woodbridge, R. M. Fyfe, et al. "The spatiotemporal spread of human migrations during the European Holocene." *Proceedings of the National Academy of Sciences USA* 117 (2020), pp. 8989–9000.

Rio, J., C. S. Quilodrán, and M. Currat. "Spatially explicit paleogenomic simulations support cohabitation with limited admixture between Bronze Age Central European populations." *Communications Biology* 4 (2021), p. 1163.

Scott, A., S. Reinhold, T. Hermes, et al. "Emergence and intensification of dairying in the Caucasus and Eurasian steppes." *Nature Ecology and Evolution* 6 (2022), pp. 813–822.

Turchin, P. "The horse bit and bridle kicked off ancient empires. A new giant dataset tracks the societal factors that drove military technology." *The Conversation*, October 21, 2021. https://theconversation.com/the-horse-bit-and-bridle-kicked-off-ancient-empires-a-new-giant-dataset-tracks-the-societal-factors-that-drove-militarytechnology-170073.

Turchin, P., D. Hoyer, A. Korotayev, et al. "Rise of the war machines: Charting the evolution of military technologies from the Neolithic to the Industrial Revolution." *PLoS One* 16 (2021), e0258161. https://doi.org/10.1371/journal.pone.0258161.

Turchin, P., H. Whitehouse, S. Gavrilets, et al. "Disentangling the evolutionary drivers of social complexity: A comprehensive test of hypotheses." *Science Advances* 8 (2022), eabn3517. https://doi.org/10.1126/sciadv.abn3517.

Vandkilde, H., S. Hansen, K. Kotsakis, et al. *Cultural Mobility in Bronze Age Europe.* BAR Publishing, 2005, pp. 5–37.

Weiss, H., M.-A. Courty, W. Wetterstrom, et al. "The genesis and collapse of Third Millennium North Mesopotamian civilization." *Science* 261 (1993), pp. 995–1004.

Wilkin, S., A. V. Miller, R. Fernandes, et al. "Dairying enabled Early Bronze Age Yamnaya steppe expansions." *Nature* 598 (2021), pp. 629–633.

Chapter 5. The Horse Before the Horse

Barkham, P. "Dutch rewilding experiment sparks backlash as thousands of animals starve." *The Guardian*, April 27, 2018. https://www.theguardian.com/environment/2018/apr/27/dutch-rewilding-experiment-backfires-as-thousands-of-animals-starve.

Bellone, R. R., H. Holl, V. Setaluri, et al. "Evidence for a retroviral insertion in TRPM1 as the cause of congenital stationary night blindness and leopard complex spotting in the horse." *PLoS One* 8 (2013), e78280. https://doi.org/10.1371/journal.pone.0078280.

Bignon, O. "De l'exploitation des chevaux aux stratégies de subsistance des Magdaléniens du Bassin parisien." *Gallia Préhistoire* 48 (2006), pp. 181–206.

Cavin, L., and N. Alvarez. *Faire revivre des espèces disparues?* Favre, 2021, p. 200.

Clottes, J. *Pourquoi l'art préhistorique?* Gallimard, 2011, p. 336.

Driscoll, D. "National parks are for native wildlife, not feral horses: Federal court." *The Conversation*, May 8, 2020. https://theconversation.com/national-parks-are -for-native-wildlife-not-feral-horses-federal-court-138204.

Fages, A., K. Hanghøj, N. Khan, et al. "Tracking five millennia of horse management with extensive ancient genome time series." *Cell* 177 (2019), pp. 1419–1435.

Fraser, M. "Britain's endangered native ponies could help habitats recover—and Brexit offers an opportunity." *The Conversation*, March 6, 2020. https://the conversation.com/britains-endangered-native-ponies-could-help-habitats -recover-and-brexit-offers-an-opportunity-122888.

Fritz, C. *L'Art de la préhistoire*. Éditions Citadelles et Mazenod, 2017, p. 626.

Guy, E. *Ce que l'art préhistorique nous dit de nos origins*. Flammarion, 2017, p. 352.

Kopnina, H. M., S. R. B. Leadbeater, and P. Cryer. "Learning to rewild: Examining the failed case of the Dutch 'New Wilderness' Oostvaardersplassen." *Communication and Education* 25-3 (2019). https://ijw.org/learning-to-rewild.

Leroi-Gouhan, A. *L'Art pariétal. Langage de la Préhistoire*. Éditions Jérome Million, 1992, p. 423.

Librado, P., C. Der Sarkissian, L. Ermini, et al. "Tracking the origins of Yakutian horses and the genetic basis for their fast adaptation to subarctic environments." *Proceedings of the National Academy of Sciences USA* 112 (2015), p. e6889– e6897.

Librado, P., N. Khan, A. Fages, et al. "The origins and spread of domestic horses from the Western Eurasian steppes." *Nature* 598 (2021), pp. 634–640.

Lovasz, L., A. Fages, and V. Amrhein. "Konik, Tarpan, European wild horse: An origin story with conservation implications." *Global Ecology and Conservation* 32 (2021), p. e01911. https://doi.org/10.1016/j.gecco.2021.e01911.

Lunt, P. H., J. L. Leigh, S. A. McNeil, and M. J. Gibb. "Using Dartmoor ponies in conservation grazing to reduce *Molinia caerulea* dominance and encourage germination of *Calluna vulgaris* in heathland vegetation on Dartmoor, UK." *Conservation Evidence Journal* 18 (2021), pp. 25–30.

Metcalf, J. L., S. J. Song, J. T. Morton, et al. "Evaluating the impact of domestication and captivity on the horse gut microbiome." *Scientific Reports* 7 (2017), p. 15497. https://doi.org/10.1038/s41598-017-15375-9.

Naundrup, P. J., and J. C. Svenning. "A geographic assessment of the global scope for rewilding with wild-living horses (*Equus ferus*)." *PLoS One* 10 (2015), e0132359. https://doi.org/10.1371/journal.pone.0132359.

Pittock, J. "Fire almost wiped out rare species in the Australian Alps. Feral horses are finishing the job." *The Conversation*, January 27, 2020. https://theconversation .com/fire-almost-wiped-out-rare-species-in-the-australian-alps-feral-horses-are -finishingthe-job-130584.

Pruvost, M., R. Bellone, N. Benecke, et al. "Genotypes of predomestic horses match phenotypes painted in Paleolithic works of cave art." *Proceedings of the National Academy of Sciences USA* 108 (2011), pp. 18626–18630.

Schubert, M., H. Jónsson, D. Chang, et al. "Prehistoric genomes reveal the genetic foundation and cost of horse domestication." *Proceedings of the National Academy of Sciences USA* 111 (2014), p. e5661–e5669.

Schubert, M., M. Mashkour, C. Gaunitz, et al. "Zonkey: A simple, accurate and sensitive pipeline to genetically identify equine F1-hybrids in archaeological assemblages." *Journal of Archaeological Science* 78 (2017), pp. 147–157.

Sommer, R. S., N. Benecke, L. Lõugas, O. Nelle, and U. Schmölcke. "Holocene survival of the wild horse in Europe: A matter of open landscape?" *Journal of Quaternary Science* 26 (2011), pp. 805–812.

Svenning, J. C. "Science for a wilder Anthropocene: Synthesis and future directions for trophic rewilding research." *Proceedings of the National Academy of Sciences USA* 113 (2016), pp. 898–906.

Walker, T. R. "Wild horses or pests? How to control free-roaming horses in Alberta." *The Conversation*, October 27, 2019. https://theconversation.com/wild-horses-or -pestshow-to-control-free-roaming-horses-in-alberta-122510.

Chapter 6. The Other Horse

Bennett, E. A., J. Weber, W. Bendhafer, et al. "The genetic identity of the earliest human-made hybrid animals, the kungas of Syro-Mesopotamia." *Science Advances* 8 (2021), eabm0218. https://doi.org/10.1126/sciadv.abm0218.

Chandezon, C. "'Il est le fils de l'âne . . .' Remarques sur les mulets dans le monde grec." In *Les Équidés dans le monde méditerranéen antique*, edited by A. Gardeisen. Actes du colloque organisé par l'École française d'Athènes, le Centre Camille-Jullian et l'UMR 5149 du CNRS, 2005, pp. 207–208.

Chappez, G. *L'Âne: histoire, mythe et réalité*. Cabedita, 2000, p. 144.

Clavel, P., J. Dumoncel, C. Der Sarkissian, et al. "Assessing the predictive taxonomic power of the bony labyrinth 3D shape in horses, donkeys and their F1-hybrids." *Journal of Archaeological Science* 131 (2021), 105383. https://doi.org/10.1016/j.jas .2021.105383.

Cucchi, T., A. Mohaseb, S. Peigné, K. Debue, L. Orlando, and M. Mashkour. "Detecting taxonomic and phylogenetic signals in equid cheek teeth: Towards new palaeontological and archaeological proxies." *Royal Society Open Science* 4 (2017), 160997. https://doi.org/10.1098/rsos.160997.

Giaimo, C. "The 1976 great American horse race was won by a mule named Lord Fauntleroy." *Atlas Obscura*, August 8, 2016. https://www.atlasobscura.com

/articles/the-1976-great-american-horse-race-was-won-by-a-mule-named-lord
-fauntleroy.

Kessler, M. "How a steeplejack, a teenager and a mule won the great American horse race." *Wbur,* September 9, 2016. https://www.wbur.org/onlyagame/2016/09/09 /virl-pierce-norton-horse-mule-race.

Lepetz, S., B. Clavel, D. Alioglu, et al. "Historical management of equine resources in France from the Iron Age to the Modern Period." *Journal of Archaeological Science: Reports* 40 (2021), 103250. https://doi.org/10.1016/j.jasrep.2021.103250.

Librado, P., and L. Orlando. "Genomics and the evolutionary history of equids." *Annual Review of Animal Biosciences* 9 (2021), pp. 81–101.

Mitchell, P. *The Donkey in Human History: An Archaeological Perspective.* Oxford University Press, 2018, p. 245.

Robinson III, C. M. "The hybrid beast that built the West." *HistoryNet,* March 28, 2018. https://www.historynet.com/hybrid-beast-built-west.

Schubert, M., M. Mashkour, C. Gaunitz, et al. "Zonkey: A simple, accurate and sensitive pipeline to genetically identify F1-equid hybrids in archaeological assemblages." *Journal of Archaeological Science* 78 (2017), pp. 147–157.

Steiner, C. C., and O. A. Ryder. "Characterization of Prdm9 in equids and sterility in mules." *PLoS One* 8 (2013), e61746. https://doi.org/10.1371/journal.pone .0061746.

Todd, E., L. Tonasso-Calvière, L. Chauvey, et al. "The genomic history and global expansion of domestic donkeys." *Science* 377 (2022), pp. 1172–1180.

Chapter 7. The Horse of the East

AbouEl Ela, N. A., K. A. El-Nesr, H. A. Ahmed, and S. A. Brooks. "Molecular detection of Severe Combined Immunodeficiency Disorder in Arabian horses in Egypt." *Journal of Equine Veterinary Science* 68 (2018), pp. 55–58.

Bernoco, D., and E. Bailey. "Frequency of the SCID gene among Arabian horses in the USA." *Animal Genetics* 29 (1998), pp. 41–42.

Brooks, S. A., N. Gabreski, D. Miller, et al. "Whole-genome SNP association in the horse: Identification of a deletion in Myosin Va responsible for Lavender Foal syndrome." *PLoS One* 6 (2010), e1000909. https://doi.org/10.1371/journal.pgen .1000909.

Cosgrove, E. J., R. Sadeghi, F. Schlamp, et al. "Genome diversity and the origin of the Arabian horse." *Scientific Reports* 10 (2020), p. 9702. https://doi.org/10.1038 /s41598-020-66232-1.

Fages, A., K. Hanghøj, N. Khan, et al. "Tracking five millennia of horse management with extensive ancient genome time series." *Cell* 177 (2019), pp. 1419–1435, e31.

Felkel, S., C. Vogl, D. Rigler, et al. "Asian horses deepen the MSY phylogeny." *Animal Genetics* 49 (2018), pp. 90–93.

Felkel, S., C. Vogl, D. Rigler, et al. "The horse Y chromosome as an informative marker for tracing sire lines." *Scientific Reports* 9 (2019), p. 6095. https://doi.org /10.1038/s41598-019-42640-w.

Fontanel, M., E. Todd, A. Drabbe, et al. "Variation in the SLC16A1 and the ACOX1 genes is associated with gallop racing performance in Arabian horses." *Journal of Equine Veterinary Science* 93 (2020), p. 103202. https://doi.org/10.1016/j.jevs .2020.103202.

Harrigan, P. "Discovery at Al-Magar. Saudi Aramco World." *Aramco World*, 2012. https://archive.aramcoworld.com/issue/201203.

Jun, J., Y. S. Cho, H. Hu, et al. "Whole genome sequence and analysis of the Marwari horse breed and its genetic origin." *BMC Genomics* 15, suppl. 9, S4 (2014). https:// doi.org/10.1186/1471-2164-15-S9-S4.

Kawai, M., Y. Minami, Y. Sayama, A. Kuwano, A. Hiraga, and H. Miyata. "Muscle fiber population and biochemical properties of whole body muscles in Thoroughbred horses." *The Anatomical Record* 292 (2009), pp. 1663–1669.

Leisson, K., Ü. Jaakma, and T. Seene. "Adaptation of equine locomotor muscle fiber types to endurance and intensive high speed training." *Journal of Equine Veterinary Science* 28 (2008), pp. 395–401.

Librado, P., C. Gamba, C. Gaunitz, et al. "Ancient genomic changes associated with domestication of the horse." *Science* 356 (2017), pp. 442–445.

Lindgren, G., N. Backström, J. Swinburne, et al. "Limited number of patrilines in horse domestication." *Nature Genetics* 36 (2004), pp. 335–336.

Mach, N., Y. Ramayo-Caldas, A. Clark, et al. "Understanding the response to endurance exercise using a systems biology approach: Combining blood metabolomics, transcriptomics and miRNomics in horses." *BMC Genomics* 18 (2017), p. 187.

Miyata, H., R. Itoh, F. Sato, N. Takebe, T. Hada, and T. Tozaki. "Effect of myostatin SNP on muscle fiber properties in male Thoroughbred horses during training period." *Journal of Physiological Science* 68 (2018), pp. 639–646.

Musial, A. D., K. Ropka-Molik, K. Piórkowska, J. Jaworska, and M. Stefaniuk-Szmukier. "ACTN3 genotype distribution across horses representing different utility types and breeds." *Molecular Biology Reports* 46 (2019), pp. 5795–5803.

Mycka, G., A. D. Musiał, M. Stefaniuk-Szmukier, K. Piórkowska, and K. Ropka-Molik. "Variability of ACOX1 gene polymorphisms across different horse breeds with regard to selection pressure." *Animals* 10 (2020), p. 2225.

Olsen, S. L. "Insight on the ancient Arabian horse from North Arabian petroglyphs." *Arabian Humanities. Revue internationale d'archéologie et de sciences sociales sur la* péninsule Arabique 8 (2017). https://kuscholarworks.ku.edu/handle/1808/27619 ?show=full.

Orlando, L., and P. Librado. "Origin and evolution of deleterious mutations in horses." *Genes (Basel)* 10 (2019), p. 649.

Pagan, J. D. "Energy and the performance horse." In *Advances in Equine Nutrition*, edited by J. D. Pagan. Nottingham University Press, 2015, pp. 141–148.

Pickering, C., and J. Kiely. "ACTN3: More than just a gene for speed." *Frontiers in Physiology* 8 (2017), p. 1080.

Remer, V., E. Bozlak, S. Felkel, et al. "Y-chromosomal insights into breeding history and sire line genealogies of Arabian horses." *Genes* 13 (2022), p. 229.

Ricard, A., C. Robert, C. Blouin, et al. "Endurance exercise ability in the horse: A trait with complex polygenic determinism." *Frontiers in Genetics* 8 (2017), p. 89.

Ropka-Molik, K., M. Stefaniuk-Szmukier, A. D. Musiał, and B. D. Velie. "The genetics of racing performance in Arabian horses." *International Journal of Genomics* (2019), 9013239. https://doi.org/10.1155/2019/9013239.

Ropka-Molik, K., M. Stefaniuk-Szmukier, T. Szmatoła, K. Piórkowska, and M. Bugno-Poniewierska. "The use of the SLC16A1 gene as a potential marker to predict race performance in Arabian horses." *BMC Genetics* 20 (2019), p. 73.

Rudolph, J. A., S. J. Spier, G. Byrns, C. V. Rojas, D. Bernoco, and E. P. Hoffman. "Periodic paralysis in Quarter Horses: A sodium channel mutation disseminated by selective breeding." *Nature Genetics* 2 (1992), pp. 144–147.

Schiettecatte, J., and A. Zouache. "The horse in Arabia and the Arabian horse: Origins, myths and realities." *Arabian Humanities. Revue internationale d'archéologie et de sciences sociales sur la péninsule Arabique* 8 (2017). https://www.academia.edu /34317012/2017_The_Horse_in_Arabia_and_the_Arabian_Horse_Origins _Myths_and_Realities?hb-g-sw=34317585.

Thomas-Derevoge, P. *Le Vizir. Le plus illustre cheval de Napoléon.* Éditions du Rocher, 2006, p. 331.

Wallner, B., N. Palmieri, C. Vogl, et al. "Y chromosome uncovers the recent Oriental origin of modern stallions." *Current Biology* 27 (2017), pp. 2029–2035.

Wutke, S., E. Sandoval-Castellanos, N. Benecke, et al. "Decline of genetic diversity in ancient domestic stallions in Europe." *Science Advances* 4 (2018), eaap9691. https://doi.org/10.1126/sciadv.aap9691.

Chapter 8. The Horse in the Middle Ages

Ameen, C., G. P. Baker, H. Benkert, et al. "Interdisciplinary approaches to the medieval warhorse." *Cheiron: The International Journal of Equine and Equestrian History* 1 (2021), pp. 100–119.

Ameen, C., H. Benkert, T. Fraser, et al. "In search of the 'great horse': A zooarchaeological assessment of horses from England (AD 300–1650)." *Journal of Osteoarchaeology* 31 (2021), p. 1247–1257.

Andersson, L. S., M. Larhammar, F. Memic, et al. "Mutations in DMRT3 affect locomotion in horses and spinal circuit function in mice." *Nature*, 488 (2012), pp. 642–646.

Barthélémy, D. *La Chevalerie*, Perrin, Tempus, 2012, p. 624.

Bouet, P. "Les chevaux de la tapisserie de Bayeux." *In Situ. Revue des patrimoines*, 27, *Le Cheval et ses patrimoines*, 2015. https://journals.openedition.org/insitu/11967.

Clavel, B., S. Lepetz, L. Chauvey, et al. "Sex in the city: Uncovering sex-specific management of equine resources from prehistoric times to the Modern Period in France." *Journal of Archaeological Science: Reports* 41 (2021), p. 103341. https://doi .org/10.1016/j.jasrep.2022.103341.

Fages, A., K. Hanghøj, N. Khan, et al. "Tracking five millennia of horse management with extensive ancient genome time series." *Cell* 177 (2019), pp. 1419–1435.

Imsland, F., K. McGowan, C.-J. Rubin, et al. "Regulatory mutations in TBX3 disrupt asymmetric hair pigmentation that underlies Dun camouflage color in horses." *Nature Genetics* 48 (2016), pp. 152–158.

Langdon, J. *Horses, Oxen and Technological Innovation: The Use of Draught Animals in English Farming from 1066–1500*. Cambridge University Press, 2002, p. 348.

Liu, X., Y. Zhang, W. Liu, et al. "A single-nucleotide mutation within the TBX3 enhancer increased body size in Chinese horses." *Current Biology* 32 (2022), pp. 480–487.

Lorans, E. *Le Cheval au Moyen Âge*. Presses universitaires François-Rabelais (Tours), 2017, p. 450.

Ludwig, A., M. Pruvost, M. Reissmann, et al. "Coat color variation at the beginning of horse domestication." *Science* 324 (2009), p. 485.

Makvandi-Nejad, S., G. E. Hoffman, J. J. Allen, et al. "Four loci explain 83% of size variation in the horse." *PLoS One* 7 (2012), e39929. https://doi.org/10.1371 /journal.pone.0039929.

Metzger, J., U. Philipp, M. S. Lopes, et al. "Analysis of copy number variants by three detection algorithms and their association with body size in horses." *BMC Genomics* 14 (2013), p. 487.

Metzger, J., R. Schrimpf, U. Philipp, and Ottmar Distl. "Expression levels of LCORL are associated with body size in horses." *PLoS One* 8 (2013), e56497. https://doi .org/10.1371/journal.pone.0056497.

Nistelberger, H. M., A. H. Pálsdóttir, B. Star, et al. "Sexing Viking Age horses from burial and non-burial sites in Iceland using ancient DNA." *Journal of Archaeological Science* 101 (2019), pp. 115–122.

Novoa-Bravo, M., K. J. Fegraeus, M. Rhodin, E. Strand, L. F. García, and G. Lindgren. "Selection on the Colombian paso horse's gaits has produced kinematic differences partly explained by the DMRT3 gene." *PLoS One* 13 (2018), e0202584. https://doi.org/10.1371/journal.pone.0202584.

Promerová, M., L. S. Andersson, R. Juras, et al. "Worldwide frequency distribution of the 'Gait keeper' mutation in the DMRT3 gene." *Animal Genetics* 45 (2014), pp. 274–282.

Regatieri, I. C., J. E. Eberth, F. Sarver, T. L. Lear, and E. Bailey. "Comparison of DMRT3 genotypes among American Saddlebred horses with reference to gait." *Animal Genetics* 47 (2016), pp. 603–605.

Ricard, A., and A. Duluard. "Genomic analysis of gaits and racing performance of the French trotter." *Journal of Animal Breeding and Genetics* 138 (2021), pp. 204–222.

Schubert, M., H. Jónsson, D. Chang, et al. "Prehistoric genomes reveal the genetic foundation and cost of horse domestication." *Proceedings of the National Academy of Sciences USA* 111 (2014), pp. e5661–e5669.

Sponenberg, P., and R. Bellone. *Equine Color Genetics.* Wiley-Blackwell, 4th Ed., 2017, p. 352.

Staiger, E. A., M. S. Almén, M. Promerová, et al. "The evolutionary history of the DMRT3 'Gait keeper' Haplotype." *Animal Genetics* 48 (2017), pp. 551–559.

Warhorse: The Archaeology of a Medieval Revolution? Accessed January 14, 2025. https://medievalwarhorse.exeter.ac.uk.

Wutke, S., L. Andersson, N. Benecke, et al. "The origin of ambling horses." *Current Biology* 26 (2016), p. R697–R699.

Wutke, S., N. Benecke, E. Sandoval-Castellanos, et al. "Spotted phenotypes in horses lost attractiveness in the Middle Ages." *Scientific Reports* 6 (2016), p. 38548.

Chapter 9. The Horse of Extreme Lands

Crubézy, É. *Vainqueurs ou vaincus? L'énigme de la Iakoutie.* Odile Jacob, 2017, p. 246.

Ferret, C. *Une civilisation du cheval. Les usages de l'équidé, de la steppe à la taiga.* Belin, 2010, p. 350.

Fuquan, Y. *The "Ancient Tea and Horse Caravan Road," the "Silk Road" of Southwest China.* Silkroad Foundation Newsletter 2, 2004. http://www.silkroadfoundation .org/newsletter/2004vol2num1/tea.htm.

Kayser, C., C. Hollard, A. Gonzalez, et al. "The ancient Yakuts: A population genetic enigma." *Philosophical Transactions of the Royal Society London B Biological Sciences* 370 (2015), 20130385. https://doi.org/10.1098/rstb.2013.0385.

Librado, P., C. Der Sarkissian, L. Ermini, et al. "Tracking the origins of Yakutian horses and the genetic basis for their fast adaptation to subarctic environments." *Proceedings of the National Academy of Sciences USA* 112 (2015), p. e6889–e6897.

Liesowka, A. "Exclusive: The first pictures of blood from a 10,000 year old Siberian woolly mammoth." *The Siberian Times*, May 29, 2013. https://siberiantimes.com /science/casestudy/news/exclusive-the-first-pictures-of-blood-from-a-10000 -year-old-siberian-woolly-mammoth/.

Liu, X., Y. Zhang, Y. Li, et al. "*EPAS1* gain-of-function mutation contributes to high-altitude adaptation in Tibetan horses." *Molecular Biology and Evolution* 36 (2019), pp. 2591–2603.

Ma, Y. F., X.-M. Han, C.-P. Huang, et al. "Population genomics analysis revealed origin and high-altitude adaptation of Tibetan pigs." *Scientific Reports* 9 (2019), p. 11463.

Nace, T. "Siberia's 'doorway to the underworld' is rapidly growing in size." *Forbes*, February 28, 2017. https://www.forbes.com/sites/trevornace/2017/02/28 /siberias-doorway-underworld-rapidly-growing-size/?sh=5bd8001b6599.

Nielsen, R., J. M. Akey, M. Jakobsson, J. K. Pritchard, S. Tishkoff, and E. Willerslev. "Tracing the peopling of the world through genomics." *Nature* 541 (2017), pp. 302–310.

Quintana-Murci, L. *Le Peuple des humains*. Odile Jacob, 2021, p. 336.

Sigley, G. "The Ancient Tea Horse Road and the politics of cultural heritage in South-west China: Regional identity in the context of a rising China." In *Cultural Heritage Politics in China*, edited by T. Blumenfield and H. Silverman. Springer, 2013, pp. 235–246.

Sigley, G. "The Ancient Tea Horse Road: The politics of cultural heritage in South-west China." *China Heritage Quarterly* 29 (2012), pp. 1–6.

Wang, G.-D., R.-X. Fan, W. Zhai, et al. "Genetic convergence in the adaptation of dogs and humans to the high-altitude environment of the Tibetan Plateau." *Genome Biology and Evolution* 6 (2014), pp. 2122–2128.

Wei, C., H. Wang, G. Liu, et al. "Genome-wide analysis reveals adaptation to high altitudes in Tibetan sheep." *Scientific Reports* 6 (2016), p. 26770. https://doi.org /10.1038/srep26770.

Wu, D.-D., C.-P. Yang, M.-S. Wang, et al. "Convergent genomic signatures of high-altitude adaptation among domestic animals." *National Science Review* 7 (2020), pp. 952–963.

Zhang, W., Z. Fan, E. Han, et al. "Hypoxia adaptations in the grey wolf (*Canis lupus chanco*) from Qinghai-Tibet plateau." *PLoS Genetics* 10 (2013), p. e1004466. https://doi.org/10.1371/journal.pgen.1004466.

Zhang, X., K. E. Witt, M. M. Bañuelos, et al. "The history and evolution of the Denisovan-EPAS1 haplotype in Tibetans." *Proceedings of the National Academy of Sciences USA* 118 (2021), p. e2020803118. https://doi.org/10.1073/pnas .2020803118.

Chapter 10. The Horse of the Americas

Collin, Y. R. H. *The Relationship between the Indigenous Peoples of the Americas and the Horse: Deconstructing a Eurocentric Myth*. PhD dissertation, University of Alaska Fairbanks, 2017. https://scholarworks.alaska.edu/handle/11122/7592.

Der Sarkissian, C., J. T. Vilstrup, M. Schubert, et al. "Mitochondrial genomes reveal the extinct *Hippidion* as an outgroup to all living equids." *Biology Letters* 11 (2015), 20141058. https://doi.org/10.1098/rsbl.2014.1058.

Feagans, C. "Pseudoarchaeological claims of Horses in the Americas," *Archaeology Reviews*, July 16, 2019. https://ahotcupofjoe.net/2019/07/pseudoarchaeological -claims-of-horses-in-the-americas.

Forrest, S. *The Age of the Horse: An Equine Journey through Human History*. Atlantic Monthly Press, 2017, p. 432.

Haile, J., D. G. Froese, R. D. E. MacPhee, et al. "Ancient DNA reveals late survival of mammoth and horse in Interior Alaska." *Proceedings of the National Academy of Sciences USA* 106 (2009), pp. 22352–22357.

Hämäläinen, P. *Lakota America: A New History of Indigenous Power*. Yale University Press, 2019, p. 544.

Hämäläinen, P. *L'Empire Comanche*. Anacharsis Éditions, 2012, p. 599.

Heintzman, P. D., G. D. Zazula, R. D. E. MacPhee, et al. "A new genus of horse from Pleistocene North America." *eLife* 6 (2017), p. e29944. https://doi.org/10.7554 /eLife.29944.

Lorenzen, E. D., D. Nogués-Bravo, L. Orlando, et al. "Species-specific responses of Late Quaternary megafauna to climate and humans." *Nature* 479 (2011), pp. 359–364.

Macfadden, B. J. "Evolution. Fossil horses—evidence for evolution," *Science* 307 (2005), pp. 1728–1730.

Mitchell, P. *Horse Nations: The Worldwide Impact of the Horse on Indigenous Societies Post-1492*. Oxford University Press, 2015, p. 496.

Murchie, T. J., A. J. Monteath, M. E. Mahony, et al. "Collapse of the mammoth-steppe in central Yukon as revealed by ancient environmental DNA." *Nature Communications* 12 (2021), p. 7120. https://doi.org/10.1038/s41467-021-27439-6.

Orlando, L. *L'ADN fossile, une machine à remonter le temps*. Odile Jacob, 2020, p. 252.

Orlando, L., A. Ginolhac, G. Zhang, et al. "Recalibrating *Equus* evolution using the genome sequence of an early Middle Pleistocene horse." *Nature* 499 (2013), pp. 74–78.

Orlando, L., C. Stépanoff, and H. Roche. "Animaux sauvages et animaux domestiques, des concepts indépassables?" In *L'Animal désanthropisé. Interroger et redéfinir les concepts*, edited by É. Baratay. Éditions de la Sorbonne, 2021, p. 320.

Raff, J. *Origin: A Genetic History of the Americas.* Twelve, 2022, p. 368.

Sacred Way Sanctuary. Accessed January 14, 2025. https://www.sacredwaysanctuary .org.

Shapiro, B. *How to Clone a Mammoth: The Science of De-Extinction.* Princeton University Press, 2015, p. 256.

Taylor, W. T. T., P. Librado, M. H. Tašunke Icu, et al. "Early dispersal of domestic horses in the Great Plains and Northern Rockies." *Science* 379 (2023), pp. 1316–1323.

Vershinina, A. O., P. D. Heintzman, D. G. Froese, et al. "Ancient horse genomes reveal the timing and extent of dispersals across the Bering Land Bridge." *Molecular Ecology* 30 (2021), pp. 6144–6161.

Wang, Y., M. W. Pedersen, I. G. Alsos, et al. "Late Quaternary dynamics of Arctic biota from ancient environmental genomics." *Nature* 600 (2021), pp. 86–92.

Willerslev, E., J. Davison, M. Moora, et al. "Fifty thousand years of Arctic vegetation and megafaunal diet." *Nature* 506 (2014), pp. 47–51.

Willerslev, E., A. J. Hansen, J. Binladen, et al. "Diverse plant and animal genetic records from Holocene and Pleistocene sediments." *Science* 300 (2003), pp. 791–795.

Chapter 11. The English Thoroughbred

"Epsom Derby winner Anthony Van Dyck dies following Melbourne Cup injury." *Horsetalk*, November 3, 2020. https://www.horsetalk.co.nz/2020/11/03 /anthony-van-dyck-dies-melbourne-cup-injury.

"Tendon injury shatters Triple Crown dream for SoCal colt I'll Have Another." CBS, June 9, 2012. https://www.cbsnews.com/sanfrancisco/news/tendon-injury -shatters-triple-crown-dream-for-socal-colt-ill-have-another/.

Bailey, E., J. L. Petersen, and T. S. Kalbfleisch. "Genetics of Thoroughbred racehorse performance." *Annual Review of Animal Biosciences* 10 (2021), pp. 131–150.

Baron, E., M. S. Lopes, D. Mendonça, and A. da Camara Machado. "SNP identification and polymorphism analysis in exon 2 of the horse myostatin gene." *Animal Genetics* 43 (2012), pp. 229–232.

Bayer, B. "The search for Ferdinand." *The Blood-Horse*, December 16, 2003. https:// www.bloodhorse.com/horse-racing/articles/178402/the-search-for -ferdinand.

Beerts, C., M. Suls, S. Y. Broeckx, et al. "Tenogenically induced allogeneic Peripheral Blood Mesenchymal stem cells in allogeneic platelet-rich plasma: 2-Year follow-up after tendon or ligament treatment in horses." *Frontiers in Veterinary Science* 4 (2017), p. 158.

Blaineau, A. *Xénophon. L'intégrale de l'oeuvre équestre.* Actes Sud, 2011, p. 280.

Bower, M. A., B. A. McGivney, M. G. Campana, et al. "The genetic origin and history of speed in the Thoroughbred racehorse." *Nature Communications* 3 (2012), p. 643.

Cash, M. M. "The death of Kentucky Derby winner Medina Spirit was at least the 75th under Bob Baffert and illuminates the dark underbelly of horse racing." *Insider*, 2021. https://www.businessinsider.com/medina-spirit-kentucky-derby -winner-dies-horse-racing-abuses-analyzed-2021-12.

Catton, P., and G. Wezerek. "Nearly half the Kentucky Derby field is racing against half-brother." *FiveThirtyEight*, May 4, 2018. https://fivethirtyeight.com/features /nearly-half-the-kentucky-derby-field-is-racing-against-a-half-brother.

Cheung, H. W., K.-S. Wong, V. Y. C. Lin, T. S. M. Wan, and E. N. M. Ho. "A duplex qPCR assay for human erythropoietin (EPO) transgene to control gene doping in horses." *Drug Testing and Analysis* 13 (2021), pp. 113–121.

Chung, M. J., S. Y. Park, J. Y. Sono, et al. "Differentiation of equine induced pluripotent stem cells into mesenchymal lineage for therapeutic use." *Cell Cycle* 18 (2019), p. 21.

De Mattos Carvalho, A., P. R. Badial, L. E. Cisneros Álvarez, et al. "Equine tendonitis therapy using mesenchymal stem cells and platelet concentrates: A randomized controlled trial." *Stem Cell Research and Therapy* 4 (2013), p. 85.

Drape, J. "Medina Spirit was pulled by the forelegs into a world that let him down." *New York Times*, May 6, 2022. https://www.nytimes.com/2022/05/06/sports /horse-racing/medina-spirit-kentucky-derby.html.

Evans, D. "Horses for courses: The science behind Melbourne Cup winners." *The Conversation*, October 30, 2014. https://theconversation.com/horses-for-courses -the-sciencebehind-melbourne-cup-winners-33362.

Fenner, K., amd M. L. Hyde. "Who's responsible for the slaughtered ex-racehorses, and what can be done?" *The Conversation*, October 20, 2019. https://thecon versation.com/whos-responsible-for-the-slaughtered-ex-racehorses-and-what -can-be-done-125551.

Fobar, R. "Why horse racing is so dangerous." *National Geographic*, January 21, 2020. https://www.nationalgeographic.com/animals/article/horse-racing-risks -deathssport.

Garcia-Roberts, G., and S. Rich. "The dark side of Bob Baffert's reign." *Washington Post*, June 18, 2021. https://www.washingtonpost.com/sports/2021/06/18/bob -bafferthorse-deaths-drug-violations.

Guest, D. J., M. R. W. Smith, and W. R. Allen. "Monitoring the fate of autologous and allogeneic mesenchymal progenitor cells injected into the superficial digital flexor tendon of horses: Preliminary study." *Equine Veterinary Journal* 40 (2008), pp. 178–181.

Henshall, C., and P. McGreevy. "Breeding Thoroughbreds is far from natural in the race for a winner." *The Conversation*, July 31, 2019. https://theconversation.com /breeding-thoroughbreds-is-far-from-natural-in-the-race-for-a-winner-121087.

Hill, E. W., R. G. Fonseca, B. A. McGivney, J. Gu, D. E. MacHugh, and L. M. Katz. "MSTN genotype (g.66493737C/T) association with speed indices in Thoroughbred racehorses." *Journal of Applied Physiology* 112 (2012), pp. 86–90.

Hill, E. W., M. A. Stoffel, B. A. McGivney, D. E. MacHugh, and J. M. Pemberton. "Inbreeding depression and the probability of racing in the Thoroughbred horse." *Proceedings of the Royal Society Biological Sciences* 289 (2022), 20220487. https://doi.org/10.1098/rspb.2022.0487.

Hogg, R. "Racing 2-year-old horses is lucrative, but is it worth the risks?" *The Conversation*, January 21, 2021. https://theconversation.com/racing-2-year-oldhorses-is-lucrative-but-is-it-worth-the-risks-152228.

Jiang, Z., J. Haughan, K. L. Moss, D. Stefanovski, K. F. Ortved, and M. A. Robinson. "A quantitative PCR screening method for adeno-associated viral vector 2-mediated gene doping." *Drug Testing and Analysis* 14 (2021), pp. 963–972.

Johnson, B. J. "Causes of death in racehorses over a 2 year period." *Equine Veterinary Journal* 26 (1994), pp. 327–330.

Maniego, J., B. Pesko, J. Habershon-Butcher, et al. "Screening for gene doping transgenes in horses *via* the use of massively parallel sequencing." *Gene Therapy* 29 (2021), pp. 236–246.

Marx, W. "Danger out of the gate." *ABC News*, February 8, 2007. https://abcnews.go.com/Sports/story?id=2857650&page=1.

Marycz, K., A. Pielok, and K. Kornicka-Garbowska. "Equine hoof stem progenitor cells (HPC) CD29+/Nestin+/K15+. A novel dermal/epidermal stem cell population with a potential critical role for laminitis treatment." *Stem Cell Reviews and Reports* 17 (2021), pp. 1478–1485.

McGill, Thomas R. Jr. "Northern Dancer, one of racing's great sires, is dead." *New York Times*, November 17, 1990. https://www.nytimes.com/1990/11/17/sports/horse-racingnorthern-dancer-one-of-racing-s-great-sires-is-dead.html.

McGivney, B. A., J. A. Browne, R. G. Fonseca, et al. "MSTN genotypes in Thoroughbred horses influence skeletal muscle gene expression and racetrack performance." *Animal Genetics* 43 (2012), pp. 810–812.

Minetti, A. E., L. P. Ardigo, E. Reinach, and F. Saibene. "The relationship between mechanical work and energy expenditure of locomotion in horses." *Journal of Experimental Biology* 202 (1999), pp. 2329–2338.

Nathanson, M. "Breeders' Cup, the Super Bowl of racing, marred by another horse's death at Santa Anita." *ABC News*, November 3, 2019. https://abcnews.go.com/US/breeders-cup-super-bowl-racing-marred-horses-death/story?id=66720756.

O'Meara, B. "The triumph and tragedy of Barbaro's fateful Triple Crown Run, 10 years later." *Bleacher Report*, May 19, 2016. https://bleacherreport.com/articles

/2640699-thetriumph-and-tragedy-of-barbaros-fateful-triple-crown-run-10
-years-later.

Okito, K., and S. Yamanaka. "Induced pluripotent stem cells: Opportunities and challenges." *Philosophical Transactions of the Royal Society London B Biological Sciences* 366 (2011), pp. 2198–2207.

Paulick, R. "Death of a Derby winner: Slaughterhouse likely fate for Ferdinand." *The Blood Horse,* July 25, 2003. https://www.bloodhorse.com/horse-racing/articles /180859/death-of-a-derby-winner-slaughterhouse-likely-fate-for-ferdinand.

Przadka, P., K. Buczak, E. Frejlich, L. Gasior, K. Suliga, and Z. Kiełbowicz. "The role of mesenchymal stem cells (MSCs) in veterinary medicine and their use in musculoskeletal disorders." *Biomolecules* 11 (2021), p. 1141.

Ribitsch, I., G. L. Oreff, and F. Jenner. "Regenerative medicine for equine musculoskeletal diseases." *Animals* 11 (2021), p. 234.

Romero, A., L. Barrachina, B. Ranera, et al. "Comparison of autologous bone marrow and adipose tissue derived mesenchymal stem cells, and platelet rich plasma, for treating surgically induced lesions of the equine superficial digital flexor tendon." *The Veterinary Journal* 224 (2017), pp. 76–84.

Rooney, M. F., E. W. Hill, V. P. Kelly, and R. K. Porter. "The 'speed gene' effect of myostatin arises in Thoroughbred horses due to a promoter proximal SINE insertion." *PLoS One* 13 (2018), e0205664. https://doi.org/10.1371/journal.pone .0205664.

Russel, A. "Is the Melbourne Cup still the race that stops the nation—or are we saying #nuptothecup?" *The Conversation,* November 1, 2021. https://theconversation .com/is-the-melbourne-cup-still-the-race-that-stops-the-nation-or-are-we -saying-nuptothecup-170801.

Smith, R. K. W., and P. M. Webbon. "Harnessing the stem cell for the treatment of tendon injuries: Heralding a new dawn?" *British Journal of Sports Medicine* 39 (2005), pp. 582–584.

Todd, E. T., S. Y. W. Ho, P. C. Thomson, R. A. Ang, B. D. Velie, and N. A. Hamilton. "Founder-specific inbreeding depression affects racing performance in Thoroughbred horses." *Scientific Reports* 8 (2018), p. 6167.

Todd, E. T., P. C. Thomson, N. A. Hamilton, et al. "A genome-wide scan for candidate lethal variants in Thoroughbred horses." *Scientific Reports* 10 (2020), p. 13153. https://doi.org/10.1038/s41598-020-69846-8.

Tozaki, T., and N. A. Hamilton. "Control of gene doping in human and horse sports." *Gene Therapy* 29 (2021), pp. 107–112.

Tozaki, T., E. W. Hill, K. Hirota, et al. "A cohort study of racing performance in Japanese Thoroughbred racehorses using genome information on ECA18." *Animal Genetics* 43 (2012), pp. 42–52.

Tozaki, T., A. Ohnuma, N. A. Hamilton, et al. "Low-copy transgene detection using nested digital polymerase chain reaction for gene-doping control." *Drug Testing and Analysis* 14 (2021), pp. 382–387.

Tozaki, T., A. Ohnuma, M. Kikuchi, et al. "Microfluidic quantitative PCR detection of 12 transgenes from horse plasma for gene doping control." *Genes (Basel)* 11 (2020), p. 457.

Tozaki, T., A. Ohnuma, M. Kikuchi, et al. "Robustness of digital PCR and real-time PCR against inhibitors in transgene detection for gene doping control in equestrian sports." *Drug Testing and Analysis* 13 (2021), pp. 1768–1775.

Tozaki, T., A. Ohnuma, M. Kikuchi, et al. "Simulated validation of intron-less transgene detection using DELLY for gene-doping control in horse sports." *Animal Genetics* 52 (2021), pp. 759–761.

Tozaki, T., A. Ohnuma, M. Kikuchi, et al. "Whole-genome resequencing using genomic DNA extracted from horsehair roots for gene-doping control in horse sports." *Journal of Equine Science* 31 (2020), pp. 75–83.

Tozaki, T., A. Ohnuma, M. Takasu, et al. "Droplet digital PCR detection of the erythropoietin transgene from horse plasma and urine for gene-doping control." *Genes (Basel)* 10 (2019), p. 243.

Tozaki, T., A. Ohnuma, M. Takasu, K. Nakamura, M. Kikuchi, and T. Ishige. "Detection of non-targeted transgenes by whole-genome resequencing for gene-doping control." *Gene Therapy* 28 (2021), pp. 199–205.

Wilkin, T., A. Baoutina, and N. Hamilton. "Equine performance genes and the future of doping in horse-racing." *Drug Testing and Analysis* 9 (2017), pp. 1456–1471.

Yamanaka, S. "From genomics to gene therapy: Induced pluripotent stem cells meet genome editing." *Annual Review of Genetics* 49 (2015), pp. 47–70.

Chapter 12. The Horse of the Future

Adli, M. "The CRISPR tool kit for genome editing and beyond." *Nature Communications* 9 (2018), p. 1911.

Aggeler, M. "Can this company take pet cloning mainstream?" *Texas Monthly*, November 2021. https://www.texasmonthly.com/news-politics/advances-in-pet-cloning.

Cohen, H. "How champion-pony clones have transformed the game of polo." *Vanity Fair*, July 2015. https://www.vanityfair.com/news/2015/07/polo-horse-cloning-adolfocambiaso.

Cohen, J. "Six cloned horses help rider win prestigious polo match." *Science*, December 13 2016. https://www.science.org/content/article/six-cloned-horses-help-rider-win-prestigious-polo-match.

Der Sarkissian, C., L. Ermini, M. Schubert, et al. "Evolutionary genomics and conservation of the endangered Przewalski's horse." *Current Biology* 25 (2015), pp. 2577–2583.

Doña Sofía Polo. Accessed January 14, 2025. https://www.dsofiapolo.com/.

Evans, M. "An inside look at equine cloning." *Horse Journals,* 2017. https://www.horse-journals.com/horse-care/alternative-therapies/inside-look-equine-cloning.

Galli, C., I. Lagutina, G. Crotti, et al. "A cloned horse born to its dam twin." *Nature* 424 (2003), pp. 635–636.

Gambini, A., O. Briski, and N. G. Canel. "State of the art of nuclear transfer technologies for assisting mammalian reproduction." *Molecular Reproduction and Development* 89 (2022), pp. 230–242.

Gambini, A., and M. Maserati. "A journey through horse cloning." *Reproduction, Fertility and Development* 30 (2018), pp. 8–17.

Hinrichs, K. "A review of cloning in the horse." *American Association Equine Practitioners Proceedings* 52 (2006), pp. 398–401.

Kraemer, D. C. "A history of equine embryo transfer and related technologies." *Journal of Equine Veterinary Science* 33 (2013), pp. 305–308.

Laffaye, H. A. *The Evolution of Polo.* McFarland & Company, 2009, p. 364.

Los Pingos del Taita. Accessed January 14, 2025. http://www.en.lospingosdeltaita.com/.

Moro, L. N., D. L. Viale, J. I. Bastón, et al. "Generation of myostatin edited horse embryos using CRISPR/Cas9 technology and somatic cell nuclear transfer." *Scientific Reports* 10 (2020), p. 15587.

Olivera, R., L. N. Moro, R. Jordan, et al. "*In vitro* and *in vivo* development of horse clone embryos generated with iPSCs, mesenchymal stromal cells and fetal or adult fibroblast as nuclear donors." *PLoS One* 11 (2016), e0164049. https://doi.org/10.1371/journal.pone.0164049.

Pilcher, H. *Life Changing: How Humans Are Altering Life on Earth.* Bloomsbury Sigma, 2020, p. 383.

"Rare horse cloned from cells taken from a stallion in 1980." *CBS News,* October 15, 2020. https://www.cbsnews.com/news/rare-horse-cloned-from-cells-taken-from-a-stallion-in-1980/.

Starr, M. "Scientists clone an endangered Przewalski's horse for the first time, and it's so cute." *Science Alert,* September 7, 2020. https://www.sciencealert.com/rare-endangered-adorable-baby-horse-is-the-first-clone-of-his-kind.

Summers, P. M., A. M. Shephard, J. K. Hodges, J. Kydd, M. S. Boyle, and W. R. Allen. "Successful transfer of the embryos of Przewalski's horses (*Equus przewalskii*) and Grant's zebra (*E. burchelli*) to domestic mares (*E. caballus*)." *Journal of Reproduction and Fertility* 80 (1987), pp. 13–20.

US Food and Drug Administration. "A primer on cloning and its use in livestock operations." FDA, 2021 .https://www.fda.gov/animal-veterinary/animal-cloning/primer-cloning-and-its-use-livestock-operations.

Usborne, D. "Polo cloning is set revolutionize the sport at Argentina's Palermo Open." *The Independent*, October 30, 2015. https://www.independent.co.uk/news /world/americas/polo-pony-cloning-is-set-revolutionise-the-sport-at-argentina -s-palermoopen-a6715646.html.

Woods, G. L., K. L. White, D. K. Vanderwall, et al. "A mule cloned from fetal cells by nuclear transfer." *Science* 301 (2003), p. 1063.

Chapter 13. Epilogue

Brooks, S. A. "Genomics in the horse industry: Discovering new questions at every turn." *Journal of Equine Veterinary Science* 100 (2021), p. 103456. https://doi.org /10.1016/j.jevs.2021.103456.

DeLuca, A. N. "World's toughest horse race retraces Genghis Khan's postal route." *National Geographic*, August 7, 2014. https://www.nationalgeographic.com/travel /article/140806-mongolia-derby-horses-genghis-riders-adventure-race.

ERC Horsepower. Accessed January 14, 2025. https://www.horsepowerproject.org.

ERC PEGASUS. Accessed January 14, 2025. https://orlandoludovic.wixsite.com /pegasus-erc.

Goullart, P. *Forgotten Kingdom*. Readers Union, John Murray, 1957, p. 259.

Orlando, L., V. Eisenmann, F. Reynier, P. Sondaar, and C. Hänni. "Morphological convergence in *Hippidion* and *Equus* (*Amerhippus*): South American equids elucidated by ancient DNA analysis." *Journal of Molecular Evolution* 57, suppl. 1 (2003), p. S29–S40.

Orlando, L., M. Mashkour, A. Burke, C. J. Douady, V. Eisenmann, and C. Hänni. "Geographic distribution of an extinct equid (*Equus hydruntinus*: *Mammalia*, *Equidae*) revealed by morphological and genetical analyses of fossils." *Molecular Ecology* 15 (2006), pp. 2083–2093.

Orlando, L., A. Ginolhac, G. Zhang, et al. "Recalibrating *Equus* evolution using the genome sequence of an early Middle Pleistocene horse." *Nature* 499 (2013), pp. 74–78.

Rousseau, É., and Y. Le Bris. *Tous les chevaux du monde. Près de 570 races et types décrits et illustrés*. Delachaux et Niestlé, 2014, p. 544 [English-language edition: *Horses of the World*, trans. Teresa Lavender Fagan. Princeton University Press, 2017].

Simpson, G. G. *Tempo and Mode in Evolution*. Columbia University Press, 1944, p. 237.

Further Reading

Chamberlin, J. E. *Horse: How the Horse Has Shaped Civilizations*. Bluebridge, 2006.

Digard J.-P. *Une histoire du cheval: arts, techniques, société*. Actes Sud, 2004.

Gouraud, J.-L., M. Woronoff, H.-P. Francfort, et al. *The Horse: From Cave Paintings to Modern Art.* Abbeville Press, 2010.

Hyland, A. *The Horse in the Ancient World.* Praeger, 2003.

Leblanc, M. A. *L'Esprit du cheval. Introduction à l'éthologie cognitive du cheval.* Belin, 2010.

Roche, D. *Culture équestre de l'Occident XVIe-XIXe.siècle. L'ombre du cheval. Le cheval moteur.* Fayard, 2008.

Roche, D. *Culture équestre de l'Occident XVIe-XIXe siècle. La gloire et la puissance.* Fayard, 2008.

Roche, D. *Culture équestre de l'Occident XVIe-XIXe siècle. Connaissances et passion.* Fayard, 2015.

Willekes, C. *The Horse in the Ancient World. From Bucephalus to the Hippodrome.* IB Tauris, 2016.

Williams, W. *The Horse: A Biography of Our Noble Companion.* Oneworld Publications, 2015.

INDEX

130, 132; of tarpans, 32–33, 35; of Yakutian horse, 150

skeletons. *See* bones

skin. *See* coat variations

SLC26A1 protein, 109–10

Snaafi Dancer (horse), 182

snake venom, 177

social anthropology, 154

social behavior of horses, 73–74, 78

social media, 184

Sondaar, Paul, 218

Sosnovka site, 29, 59–60

Spain. *See* Iberian Peninsula

specimens. *See* biological tissue

speed: breeding for, 189–94; of medieval horses, 132; physiology of, 110–11; sprint racing (*see* racing)

spine, of horses, 44–47, 126–27

spoked wheels, 60–62

sponges, stuffing horse's nostrils with, 176

sporting industry. *See* polo horses; racing

spotted coats, 125–26

sprint racing. *See* racing

St. Petersburg Zoological Museum (Russia), 208

stallions: Arabian, 104–6; in breeding herds, 15; and breeding in racing industry, 180–81, 183, 197–98; cloning of, 199–200, 212; and lineage testing, 43; in rural versus urban settings, 138; sacrifice of, 129–30; social behavior of, 73–74

Standard of Ur, 98

stearic acids, 18

steed (medieval warhorse), 130–33, 137, 139

stem cells, 185–87

stenonines, 156

stirrups, 65, 131

stone carvings, 100–102

strength: of medieval horses, 132–33; of mules, 88

strontium, 165

substitutions. *See under* mutations

sugars (glycogen), 108–11, 193

Sumerian culture, 98

Šuŋkawakaŋ Nation, 167–70, 223

surrogate mares, 202–3

Sweden, 128–29

Switzerland, 135

S'yezzhe burial site, 30

symbolic role of horses: in art (*see* artwork); climate argument and, 35; funerary (*see* funerary rituals); during Middle Ages, 119–21; in religion (*see* religion, horse in)

Syria, 118, 191

Syrian onager, 98

tail of horse, language related to, 51

TAKH association, 74

tarpan, 27–32, 223; archeological studies of, 29–32; breeding of, 32–33; climate suitable for, 35; extinction of, 79; genomic studies of, 79–80; size of, 32–33, 35

Tavan Tolgoi site, 129

Taylor, Will, 164–65, 168

TBX3 gene, 136–37

Tea Horse Road, 140–41, 143, 223

teeth: effect of bits on, 12–13, 31–32; individual characteristics identified by, 73–74, 139; isotopic studies of, 165

temperament. *See* behavioral characteristics

tendon wounds, 184–87

A NOTE ON THE TYPE

This book has been composed in Arno, an Old-style serif typeface in the classic Venetian tradition, designed by Robert Slimbach at Adobe.